Animal Breeding and Genetics

NIPA® GENX ELECTRONIC RESOURCES & SOLUTIONS P. LTD.
New Delhi-110 034

www.nipaers.com

Browse, Search, Read & Buy...

Print Books eChapters
eBooks   Publishers
Forthcoming Subject Catalogues
New Titles

Online Resources on:
- ✓ Current Affairs
- ✓ Reasoning, Logic & Aptitude
- ✓ Competitive Examinations
- ✓ English Language Lab
- ✓ Writing & Pronunciation Tools
- ✓ Personality Development
- ✓ Online Programmes for Professionals
- ✓ Interview Preparation

Pay Using

 PhonePe   Google Pay   paytm   PayPal   UPI   CC-Avenue

Animal Breeding and Genetics

C.V. Singh, Professor and Head

Department of Animal Genetics and breeding

College of Veterinary and Animal Sciences

G.B. Pant University of Agriculture and Technology

Pantnagar-263145, Uttarakhand

NIPA® GENX ELECTRONIC RESOURCES & SOLUTIONS P. LTD.
New Delhi-110 034

**NIPA® GENX ELECTRONIC
RESOURCES & SOLUTIONS P. LTD.**

101,103, Vikas Surya Plaza, CU Block
L.S.C. Market, Pitam Pura, New Delhi-110 034
Ph : +91 11 27341616, 27341717, 27341718
E-mail: newindiapublishingagency@gmail.com
www: www.nipabooks.com

For customer assistance, please contact
Phone: + 91-11-27 34 17 17
Fax: + 91-11- 27 34 16 16
E-Mail: feedbacks@nipabooks.com

© 2023, Publisher

ISBN: 978-81-19002-05-4

All rights reserved. No part of this publication may be reproduced, stored in a retrieval system or transmitted in any form or by any means, including electronic, mechanical, photocopying recording or otherwise without the prior written permission of the publisher or the copyright holder.

This book contains information obtained from authentic and highly reliable sources. Reasonable efforts have been made to publish reliable data and information, but the author/s, editor/s and publisher cannot assume responsibility for the validity, accuracy or completeness of all materials or information published herein or the consequences of their use. The work is published with the understanding that the publisher and author/s are not attempting to render any professional services. The author/s, editor/s and publisher have attempted to trace and acknowledge the copyright holders of all material reproduced in this publication and apologize to copyright holders if permission and/or acknowledgements to publish in this form have not been taken. If any copyrighted material has not been acknowledged, please write to us and let us know so that we may rectify the error, in subsequent reprints.

Trademark Notice: NIPA®, the NIPA® logos and their presentations (the way they are written/ presented) in this book are the trademarks of the publisher and hence may not be used without written permission, if copied or used without authorization, the infringer will be prosecuted as per law.

NIPA® also publishes books in a variety of electronic formats. Some content that appears in print may not be available in electronic books, and vice versa.

Composed and Designed by NIPA®.

Preface

Livestock production occupies an important place in agriculture economy of India both through contribution of GDP and employment especially in rural and peri-urban areas. It is also contributing highly nutritious protein food at relatively cheaper prices for human population.The main objective of animal breeding is to bring improvement in livestock using different traits of economic importance. Majority of the economic traits are controlled by multiple genes mostly additively and thus are quantitative in nature rather than being quantitative. The basic principles of inheritance and breeding are the same for all animals; however, implementation of breeding schemes differs considerably from species to species. Reasons for this include differences in reproductive capacity, potential to record traits of interest and available resources for research and implementation. Hence, this book attempts to describe applied breeding methods for different domestic animal species as currently implemented. However, an attempt has been made to include international perspectives wherever relevant because students are expected to deal with international challenges in their future careers. As far as possible, the chapters have been structured the same way for easy across-species comparisons.

In this book, brief history of population genetics, **domestication of livestock species, classification of breeds**, economic characteristics of different livestock species & poultry and their importance, basic statistics, qualitative and quantitative inheritance, gene and genotype frequency and factors influencing gene frequency, values and means of population, methods of estimation and uses of heritability and repeatability, correlations, selection, response to selection, basis of selection, progeny testing, open nucleus breeding system, sire evaluation ,methods of selection, breeding or mating systems, heterosis or hybrid vigor definitions and **current livestock and poultry breeding programmes** have been discussed in different chapters. The subject matter has been dealt with in a logical sequence so that the reader is conveyed from simple to more complex interpretation with relative ease. It is felt that the reader which are likely to comprise mostly of graduate and post graduate student of animal breeding and researcher will be able to get a deeper insight and better perceptions into the realm of the dynamic science of animal breeding.

—C.V.Singh

Contents

1. **Brief History of Population Genetics** 1
   Introduction 1
   Definition of Animal Breeding 3
   Beginnings 4

2. **Domestication of Livestock Species** 9
   Domestication of Livestock 9
   Flexible Diet 11

3. **Classification of Livestock Breeds** 21
   Classification of Breeds 21
   The Bos Taurus is again Divided into Three Subgroups. 21
   Definition of Breed 22
   Types of Breeds 22
   Classification of Cattle Breeds 22
   Milch Breeds 23
   Dual Purpose Breeds 26
   Draught Breeds 28
   Exotic Dairy Cattle Breeds 40
   Crossbred Cattle Strains 42
   Synthetic Exotic Cattle Breeds Evolved 44
   Synthetic Dairy Cattle Breeds Evolved in India 45
   Classification of Buffalo Breeds of India 46
   Classification of Sheep Breeds 56
   Exotic Sheep Breeds 70
   Synthetic Sheep Breeds Evolved in India 73
   Classification of Indian Goat Breeds 74
   Exotic Goat Breeds 98
   Classification of Pigs 100
   Classification of Fowls 102

	Common Breeds of Poultry	103
	American Class	104
	Asiatic Class	104
	Mediterranean Class	105
	English Class	106
	Indian Breeds	106
	Common Breeds of Duck	108
4.	**Economic Characters of Livestock and Poultry**	**109**
	Economic Traits of Swine	112
	Growth Rate	113
	Feed Efficiency	113
5.	**Basic Statistics**	**115**
	Introduction	115
	Most Frequently Occuring Value (mode)	115
	The Sampling Distribution and Standard Deviation of the Mean	118
	Correlation Coefficient (r value)	121
	Chi-squared Test	125
6.	**Qualitative and Quantitative Inheritance**	**129**
	Quantitative Versus Qualitative Inheritance	129
	Qualitativ Traits	130
	Models for Quantitative Inheritance	132
7.	**Gene and Genotype Frequency and Factors Influencing Gene Frequency**	**135**
	Gene Frequency	135
	Autosomal Loci with Multiple Alleles	137
	Mathematical Models of Genetic Drift	147
	Other Models of Drift	148
8.	**Variation and Measurements**	**151**
	Components of Variance	151
9.	**Values and Means of Population**	**157**

	Average Effect	159
	Breeding value	161
	Interaction	163

10. Heritability – Methods of Estimation and Use 165
    Estimation of Heritability 168
    Correlation Estimates of Heritability 171
    ANOVA 185
    Importance of Heritability 189
    Repeatability – Methods of Estimation and Uses 191

11. Correlations 201
    Genetic Correlation 201
    Importance of Genetic correlation 203
    Estimation of Genetic Correlation 203
    Phenotypic and Environmental Correlations 208
    Estimation of environmental Correlations 209
    Recision of Estimates of Genetic Correlations 210

12. Selection – Natural Vs Artificial Selection 215
    Natural Selection 215
    Artificial Selection 216
    Selection for Different kinds of Genes 217

13. Response to Selection 219
    Components of the Response 220
    Improvements of Response 225
    Generation Interval or Generation Length 226
    Measurement of Response 227

14. Basis of Selection 233
    Individual Selection 233
    Pedigree Selection 234
    Family Selection/SIB Selection 235
    Within Family Selection 236

    Progeny Testing 236
    Testing Procedure 237
    Requirements For Progeny Testing 238

15. **Methods of Selection** 239
    Tandom Method 239
    Independent Culling Level Method 239
    Index Selection 240
    General Considerations and Limitations 242

16. **Open Nucleus Breeding System** 243
    Multiple Ovulation Embryo Transfer and Open Nucleus Breeding System 245
    Nucleus Size 248
    Migration Rate 249
    Change in Genetic Variance 250
    Theoretical Results 251
    Demerits 252
    Conclusion 252

17. **Methods of Sire Evaluation** 255
    Equivalent Parent or Intermediate Index 256
    Best Linear Unbiased Predictor Method (BLUP) 260

18. **Breeding or Mating Systems** 267
    Random Mating 267
    Breeding For Increased Homozygosity 268
    Measurement of Inbreeding & Relationship 270
    Coefficient of Relationship 277
    Rotational Cross Breeding 286
    Criss Crossing 286
    Top Crossing 287
    Selection for Combining Ability 289
    Selection for General and Specific Combing Ability 290
    Recurrent Selection 290

Contents/xi

19. Heterosis or Hybrid Vigor 291
Definitions 291
Theory of Genetic Balance 292
Practical Application of Heterosis 293
Periodic Selection 295
Cumulative Selection 295
Gemete Selection 295
Retention of Heterosis 295
Types of Heterosis 300
Use of Heterosis 301
Types of Dominance 302
Loss of Heterosis 303

20. Selection of Dairy Cattle & Buffaloes 307
General Selection Procedures for Dairy Breeds 307

21. Current Livestock and Poultry Breeding Programmes 311
Cattle Breeding Policies and Programmes in the Planning Process 311
Infrastructure For Cattle Development and Breeding
Central Herd Registration Scheme 318
Central Frozen Semen Production 320
Breeding Policy in India 323
Cattle Breeding Policy in Different States 326

22. Conservation of Genetic Resources in India 331
Introduction 331
Genetic Diversity in India 332
Justification For Conservation or Reasons? 332
Mechanism of Conserving Cattle Genetic Resources 334
Ex-Situ/Cryogenic Preservation Includes 337
Ex-Sity Conservation 336

References 343

About the Author

1. Dr. C.V. Singh: Presently Professor & Head, Department of Animal Genetics and Breeding, College of Veterinary and Animal Science, G.B. Pant University of Agriculture &Technology Pantnagar, U.S.Nagar has about 30 years of teaching and research experience in Animal and Poultry Breeding and held the position of Principal, CCSSDS Degree College Iglas, Aligarh, Joint Director, Instructional Dairy/Poultry Farm of the university and some other administrative positions in the university is animal breeder of repute, who identified several new germplasm viz. Chaugarkha and Udaipuri Goats and Hill Cattle and developed strain of commercial broiler. He organized two national symposiums. He has guided more than 15 Post Graduate students in the field of Animal and Poultry Breeding. He is the author of some other books and monographs and published around 300 research papers on different aspects of animal production and breeding in peer reviewed journals of national and international repute. He is the member of many professional societies and scientific board.

CHAPTER - 1

Brief History of Population Genetics

Introduction

From the very early days human beings depend on animals and animal products for food and other requirements. In dairy and poultry farms high yielding animals are reared. These high yielding animals are produced by hybridization experiments. Previously the animals were developed basing on unscientific methods. Before the discovery of principles of heredity human beings have selected the animals with required characters and learned to develop the plants having the selected characters. This phenomenon is called Artificial selection. However, an increased knowledge of biology, especially genetics, has helped in improving the quality of animals and animal products as per the human requirements.

Animals throughout the world supply human beings with milk, meat, egg, draft power, transportation, hides, fertiliser and many other useful products. Therefore, animal breeding is the beginning or the foundation to meet out the requirement. No structure can be stronger than its foundation. Hence, it behoves agriculturists and livestock breeders especially to give special attention to their programme of animal breeding. The face of animal breeding has changed significantly over the past decades. Animal breeding used to be in the hands of a few distinguished 'breeders', individuals who seems to have specific arts and skills to 'breed good livestock'. Now-a-days, animal breeding is much dominated by science and technology. In some livestock species, animal breeding is in the hands of large companies, and the role of individual breeders seems to have decreased. There are several reasons for this change. Firstly, the breeding industry has taken up scientific principles. Looking was replaced by measuring, and an intuition was partly replaced by calculations and scientific prediction. Other major developments were caused by the introduction of biotechnology. These are roughly the reproductive technologies, and the molecular genetic technology. Not all of this is new. Artificial insemination was introduced in the fifties in cattle. No doubt that the technology had a major impact on rates on genetic improvement in dairy cattle, and just as important, on the structure of animal breeding programs. Nowadays, technologies like ovum pick up, in vitro fertilization, embryo transfer, cloning of individuals, cloning of genes, and selection with the use of DNA markers are all on the ground. Some of the technologies are already applied, others are further developed, or waiting for application. Finally, the rapid development of computer and information technology has greatly influenced data collection and genetic evaluation procedures in livestock populations, now allowing comparison of breeding values across herds, breeds or countries. The introduction of breeding methods typically needs to find the right balance between what is possible from a technological point of view and what is accepted by the decision makers and users within the socio-economic context of a production system. Ultimately it is the consumer who decides which technology is desirable or not. In most western societies, consumers are increasingly aware of health, environmental and animal welfare issues. Food safety and methods food production are part of their buying behaviour. However, price and production efficiency remain to be major contributors to sustainability of a livestock industry. Successful animal breeding programs need to find the right dose of technology that helps them to be competitive.

Animal breeding is a fascinating discipline. It has long been recognised as one branch of arts and only recently it started to be recognised as a special branch of science. It is also one of the steps in the process of animal production, but it is the first step and fundamental to a sound animal husbandry. Application of improved methods of breeding, feeding, management and disease control during the last few decades has greatly increased the efficiency of production.

Animal breeding is the application of genetics and physiology of reproduction to animal improvement. The purpose of animal breeding is not to genetically improve individual animals but to improve whole animal population i.e. to improve future generations of animals. To achieve this improvement, the breeder is provided with two important tools: Selection and Breeding. These two tools are the decision making in livestock improvement. Selection decides which animals are going to become parents to produce offspring for the future generation and breeding decides which males should be mated with which females. Therefore applying proper selection and system of breeding, the improvement in type, production, longevity, regularity of breeding etc. as well as the ability to transmit these desirable qualities to many progenies will result.

The animal breeder faces many complex problems during hybridization experiments because many traits of animals are dependent on the interaction of multiple genes. When the attempts are made only to increase the size of eggs in fowls, it was observed that the progeny produced yielded few number of eggs or even they die sometimes. That is if only one character is taken for improvement of the animals; the other characters will degenerate or result in harmful effects. Hence at the time of selection all the desirable characters are to be taken into consideration. The techniques for the improvement of animals involve principles of selection based on quantitative variations. It is not possible for all of the desirable traits to be obtained in one individual. The successful product must contain maximum number of desirable traits and a minimum number of undesirable traits.

Definition of Animal Breeding

Animal breeding, as the application of science to the genetic improvement of animals, implies a close inter relationship between theory and its application. Genetics has provided the matrix form which logical principles could be developed and tested by experimentation and practice.

During the 20th century, genetics has assumed a broader meaning and numerous subdivisions of the discipline has emerged. Advances in animal breeding also have drawn heavily on contributions from statistics, biochemistry, physiology, economics and other disciplines. Animal breeding and reproductive physiology have been uniquely interwoven.

Animals with superior genotypes can not contribute to succeeding generations unless their reproductive capacity is maintained at satisfactory levels.

Livestock production is an economic enterprise. Hence, animal breeding recommendations must withstand the scrutiny of economic, as well as genetic considerations, before they are accepted and integrated into enterprises by breeders. As a consequence, a close relationship between the breeder and the science, application and theory has been nurtured.

Beginnings

Pre – Mendelian

Even before Mendel's principles were uncovered, the mystery of the transmission of hereditary material from parent to offspring was recognized although not understood, Robert Backwell of Leicestershire, England who lived from 1725 to 1795, is generally credited as the founder of animal breeding. He contributed much to the development of the longhorn cattle, Leicester sheep and shire horses. Backwell reportedly purchased animals from different sources which he thought met his goals and utilized inbreeding and intense selection to develop the kind of animals he desired. His animals were much in demand, and he appears to have been among the first to conduct systematic progeny tests of rams and bulls.

Apparently Backwell told little about his mode of operation and some were obliged to think that there was something mysterious about it. At that time there was a strong prejudice against inbreeding, and it may have been that he did not want to invite criticism. Many stockman did come to work with Backwell, perhaps the most noted, the colling brothers, are credited with founding the short horn breed.

Herd books provided records of identification and parentage of foundation stock. The first herd book was published in 1871 for thorough bred horses. The coats shorthorn herd book appeared in 1822, accepting for registry only what were judged as outstanding animals. The herd book fever reached a peak in the 1870's and 1880's when importers particularly wanted to declare that their animals could be traced to a known foundation. Herd books and breed societies have been a port of animal breeding with nobly stated purposes, yet on occasion, hindering real genetic progress.

Practical breeding moved ahead even without the benefit of a scientific foundation. Darwin's studies of populations drew upon the experience of practical breeders. Conversely, reports suggest that the philosophy and writings of Darwin on organic evolution and natural selection were certainly known by some British breeders. Darwin explained the impact of reproductive fitness logically, but his insights into the process of hereditary transmission were flawed. Under the blending hypothesis the variation in the population would be halved each generation as pointed out by Jenkins (1967). A tremendous infusion of new variability would be required to reconcile the continuing variation observed in parmictic populations. Hence, favourable circumstances were provided for the acceptance of De Vries (1910) mutation theory.

Darwin (1859) underscored the importance of continuous variation to evolutionary and directed change in animals and plants. Galton (1889) without the benefit of Mendel's findings added quantitative precision to the characterization of the variation which Darwin had drawn upon to formulate his theory of natural selection. Galen's law stated that given a correlation of 0.50 between parent and offspring in a population with minimal inbreeding and for a highly hereditary trait, the correlation between an animal and a more remote ancestor is halved for each intervening generation. Lush (1945) pointed out that practical breeders in Britain had used this concept as early as 1815.

A second component of Galton's law was that the square of the correlation coefficient measures the proportion of the variance in one variable which disappears when the other is held constant. Yet Galton failed to recognize that this postulate was not true when bridging over generations. Each of the four grand parents do not account for 1/16 of the variance of their individual independent of the parents. It was Yule (1906) who corrected this view of Galton with his multiple regression technique.

Mendelism Rediscovered

Mendel (1865) had already provided the answer to the flaw in Darwin's theory of blending inheritance, although, as we now know, the scientific world had not at that time recognized the moment of the findings. His conclusion that the basic hereditary mechanism was particulate and not blending, with the capability for independent assortment and recombination gave substance to Darwin's theory of natural selection. According to Wright (1967) Mendel accepted the theory of Natural Selection, although neither Darwin nor Mendel recognized this in their time.

The veil of ignorance about the fundamental postulates of heredity was lifted in 1900 when De Vries in Holland, Correns in Germany and Techermak in Austria recognized the impact of Mendel's 1865 publication. A flurry of investigations moved forward to uncover examples of inheritance which followed the postulates of Mendel. Bateson was an active investigator in this area at Cambridge. His study of the inheritance of comb types in chickens brought forth one of the first and most interesting examples of gene interaction. Batson (1903) is credited with suggesting Genetics as the name for the new discipline. He also proposed the terms homozygotes and heterozygotes. Johansen (1903) made the distinction between genotypes and phenotypes and is credited with the first use of these terms as well as the designation of the gene.

Mendelism and the Biometrical Approach

Expected parent offspring and maternal correlations under Mendelism with random mating were developed by Pearson (1904), Yule (1906) postulated a swarm of genes affecting a trait, which had small but similar effects, as a contribution to reconciling Mendelism with the emerging biometrical analysis of genetic variation.

One of the comer stones of animal breeding and population genetics was revealed independently by Hardy (1908) and Weinberg (1908), which has become known as the Hardy-Weinberg law. These findings further showed how population variability was maintained in a large random mating population with the genotypic or zygotic frequencies being equal to the square of the gametic frequencies. Fisher's (1918) demonstrated that the results of the statistical analysis were embedded in the principles of Mendel.

The impact of dominance variance has been examined for the most part assuming no epistasis. Comstock and Robinson (1952) proposed experimental designs to clarify the nature of dominance effects with particular concern for evidence of over dominance. Wright (1935) provided insights into the contribution of two gene epistatic deviations to the correlation between relatives as he considered deviations from an optimum. Cockerham (1954) presented an extension of the concept of partitioning the epistatic component of variance and the contribution of the several partitions to the correlation between relatives. This drew upon the partitioning of individual degrees of freedom for a loci model, recognizing aditive effects, dominance deviations and epistatic effects.

Fisher's (1925) contribution of the analysis of variance has been one of the most useful to statistical methodology in animal breeding. Apart from its use to test mean differences, its application for the estimation of components of genetically and environmentally caused variation has been vide spread. Wright (1918) earlier had developed the path coefficient analysis to relate the correlation between variables in an inter cutting system of casual relationships. The intraclass correlation introduced by Harris (1913) has had much applicability in animal breeding. Variance ratios have been used widely with the concepts of repeatability and heritability as introduced by Lush having most common usage. Unequal and disproportionate sub class numbers, being the urle for animal data, has required special adaptations of the analysis of variance procedures (Harrey, 1975). Winson and Clark (1940) were among the first to provide expectations of mean squares for data with unequal subclass numbers, Yate's (1938) method of fitting constants was adopted for testing mean differences and Henderson (1953) expanded this approach to account for adjustment of fixed effects, plus the adaptation of maximum likelihood in variance component estimation. Henderson's (1953) methods II and III provided approaches to adjust for the fixed effects in a linear model. Subsequently methods of maximum likelihood estimation with a general mixed linear model were advanced by Hartley & Rao (1967), Patterson and Thompson (1971) and Rao (1971). Maximum likelihood and REML modifications, while biased, avoid the troubling negative estimates of components.

Till 500 A.D. when the fall of Roman Empire began animal breeding was at its esteem. With the fall of Roman Empire for about 1000 years called Dark and middle Ages, animal husbandry was at a still.

From 1700 A.D., again there was an improvement. The beginning of modern animal breeding is to be found mainly in England and Europe.

The British Royalty encouraged horse breeding especially for race horses. The Earls and Dukes imported bulls from Holland and bred their native stocks. Dutch cattle were introduced into Hereford shire that laid the foundation of the present Hereford cattle. By crossing the native and Dutch cattle and subsequent inbreeding, the British cattle were improved far beyond the best.

Landmarks of Animal Breeding in the World

1725 – 1795 Robert Bakewell, an English man began his animal breeding work at Dishley, Leicestershire, England with horses, sheep and cattle. He is called Father of Animal Breeding. He travelled extensively for his time both in England and on the continent in quest of superior breeding stock. He developed theories and tested them with experiments. He concentrated on producing farm animals with increased efficiency. Bakewell's two remarks were "Like begets like" and "Breed the best to the best".

Out of his experiments and observations he developed some general principles. They were: Has got definite ideals/objectives/goals. For example, beef cattle – a low set blocky and quick maturity.

Practiced sire testing by leasing the sires to other breeders and those that proved most satisfactory was brought back for use on his own females.

"Breed the best to the best" regardless of relation ship and this led to extremely close breeding.

Performed progeny testing of bulls and rams.

Introduced inbreeding as tool in livestock improvement.

"Like be gets like"

Bakewell's methods were widely copied and thus the foundation of purebred was laid. He laid the foundation for the Shire horses, Leghorn cattle and Leicester sheep. In 1775 Collings brothers copied the Robert Bakewell's method and laid foundation for the Shorthorn cattle. The Shorthorn bull "Comet" has an inbreeding co-efficient of 47 %.

In 1791 British Royalty encouraged horse breeding for races, which results in English thoroughbred and general studbook.

Dukes of England imported bulls from Holland and bred their native cattle – Dutch cattle were introduced into Hereford which laid the foundation for the present Hereford cattle.

Tompkins and Galliers laid the foundation for Hereford breed of cattle in England.

Chapter - 2

Domestication of Livestock Species

Domestication of Livestock

Domestication (from Latin domesticus) is the process whereby a population of animals or plants is changed at the genetic level through a process of selection, in order to accentuate traits that benefit humans. It differs from taming in that a change in the phenotypical expression and genotype of the animal occurs, whereas taming is simply the process by which animals become accustomed to human presence. In the Convention on Biological Diversity, a domesticated species is defined as a "species in which the evolutionary process has been influenced by humans to meet their needs." Therefore, a defining characteristic of domestication is artificial selection by humans. Humans have brought these populations under their control and care for a wide range of reasons: to produce food or valuable commodities (such as wool, cotton, or silk), for types of work (such as transportation, protection, and warfare), scientific research, or simply to enjoy as companions or ornaments. Plants domesticated primarily for aesthetic enjoyment in and around the home are usually called house plants or ornamentals, while those domesticated for large-scale food production are generally called crops. A distinction can be made between those domesticated plants that have been deliberately altered or selected for special desirable characteristics and those plants that are used for human benefit, but are essentially no different from the wild populations of the species. Animals domesticated for home companionship are usually called pets while those domesticated for food or work are called livestock or farm animals.

Described how the process of domestication can involve both

unconscious and methodical elements. Routine human interactions with animals and plants create selection pressures that cause adaptation as species adjust to human presence, use or cultivation. Deliberate selective breeding has also been used to create desired changes, often after initial domestication. These two forces, unconscious natural selection and methodical selective breeding, may have both played roles in the processes of domestication throughout history. Both have been described from man's perspective as processes of artificial selection. The domestication of wheat provides an example. Wild wheat falls to the ground to reseed itself when ripe, but domesticated wheat stays on the stem for easier harvesting. There is evidence that this critical change came about as a result of a random mutation near the beginning of wheat's cultivation. Wheat with this mutation was harvested and became the seed for the next crop. Therefore, without realizing, early farmers selected for this mutation, which would otherwise have died out. The result is domesticated wheat, which relies on farmers for its own reproduction and dissemination. Mutation is not the only way in which natural and artificial selection operate. Darwin describes how natural variations in individual plants and animals also support the selection of new traits. It is speculated that tamer than average wolves, less wary of humans, selected themselves as domestic dogs over many generations. These wolves were able to thrive by following humans to scavenge for food near camp fires and garbage dumps. Eventually a symbiotic relationship developed between people and these proto-dogs. The dogs fed on human food scraps, and humans found that dogs could warn them of approaching dangers, help with hunting, act as pets, provide warmth, or supplement their food supply. As this relationship progressed, humans eventually began to keep these self-tamed wolves and breed from them the types of dogs that we have today.

In recent times, selective breeding may best explain how continuing processes of domestication often work. Some of the best-known evidence of the power of selective breeding comes from an experiment by Russian scientist, Dmitri K. Belyaev, in the 1950s. His team spent many years breeding the Silver Fox (*Vulpes vulpes*) and selecting only those individuals that showed the least fear of humans. Eventually, Belyaev's team selected only those that showed the most positive response to humans. He ended up with a population of grey-coloured foxes whose behavior and appearance was significantly changed. They no longer showed any fear of humans and often wagged their tails and licked their human caretakers to show affection. These foxes had floppy ears, smaller skulls, rolled tails and other traits commonly found in dogs.

Despite the success of this experiment, it appears that selective breeding cannot always achieve domestication. Attempts to domesticate many kinds of wild animals have been unsuccessful. The zebra is one example. Despite the fact that four species of zebra can interbreed with and are part of the same genus as the horse and the donkey, attempts at domestication have failed. Factors such as social structure and ability to breed in captivity play a role in determining whether a species can be successfully domesticated. In human history to date, only a few species of large animal have been domesticated. In approximate order of their earliest domestication these are: dog, sheep, goat, pig, ox, yak, reindeer, water buffalo, horse, donkey, llama, alpaca, Bactrian camel and Arabian camel.

According to evolutionary biologist Jared Diamond, animal species must meet six criteria in order to be considered for domestication.

Flexible Diet

Creatures that are willing to consume a wide variety of food sources and can live off less cumulative food from the food pyramid (such as corn or wheat), particularly food that is not utilized by humans (such as grass and forage) are less expensive to keep in captivity. Carnivores by definition feed primarily or only on animal tissue, which requires the expenditure of many animals, though they may exploit sources of meat not utilized by humans, such as scraps and vermin.

Reasonably Fast Growth Rate

Fast maturity rate compared to the human life span allows breeding intervention and makes the animal useful within an acceptable duration of caretaking. Large animals such as elephants require many years before they reach a useful size.

Ability to be Bred in Captivity

Creatures that are reluctant to breed when kept in captivity do not produce useful offspring, and instead are limited to capture in their wild state. Creatures such as the panda, antelope and giant forest hog are territorial when breeding and cannot be maintained in crowded enclosures in captivity.

Pleasant Disposition

Large creatures that are aggressive toward humans are dangerous to keep in captivity. The African buffalo has an unpredictable nature and is highly dangerous to humans; similarly, although the American bison is raised in enclosed ranges in the Western United States, it is much too dangerous to be regarded as truly domesticated. Although similar to the domesticated pig in many ways, Africa's warthog and bush pig are also dangerous in captivity.

Temperament which makes it Unlikely to Panic

A creature with a nervous disposition is difficult to keep in captivity as it may attempt to flee whenever startled. The gazelle is very flighty and it has a powerful leap that allows it to escape an enclosed pen. Some animals, such as the domestic sheep, still have a strong tendency to panic when their flight zone is encroached upon. However, most sheep also show a flocking instinct, whereby they stay close together when pressed. Livestock with such an instinct may be herded by people and dogs.

Modifiable Social Hierarchy

Social creatures whose herds occupy overlapping ranges and recognize a hierarchy of dominance can be raised to recognize a human as the pack leader: tapirs and rhinoceroses are solitary and do not tolerate being penned with each other antelope and deer except for reindeer are territorial when breeding and live in herds only for the rest of the year bighorn sheep and peccaries have non hierarchical herd structures and do not follow any definite leader: instead males fight continuously with each other for mating opportunities. Musk ox herds (although having a defined leader) maintain mutually exclusive territories and two herds will fight if kept together.

However, this list is of limited use because it fails to take into account the profound changes that domestication has on a species. While it is true that some animals, including parrots, whales, and most members of the Carnivora, retain their wild instincts even if born in captivity, some factors must be taken into consideration.

In particular, number (5) may not be a prerequisite for

domestication, but rather a natural consequence of a species' having been domesticated. In other words, wild animals are naturally timid and flighty because they are constantly faced by predators; domestic animals do not need such a nervous disposition, as they are protected by their human owners. The same holds true for number (4) – aggressive temperament is an adaptation to the danger from predators. A Cape buffalo can kill even an attacking lion, but most modern large domestic animals were descendants of aggressive ancestors. The wild boar, ancestor of the domestic pig, is certainly renowned for its ferocity; other examples include the aurochs (ancestor of modern cattle), horse, Bactrian camels and yaks, all of which are no less dangerous than their undomesticated wild relatives such as zebras and buffalos. Others have argued that the difference lies in the ease with which breeding can improve the disposition of wild animals, a view supported by the failure to domesticate the kiang and onager. On the other hand for thousands of years humans have managed to tame dangerous species like the elephants, bears and cheetahs whose failed domestications had little to do with their aggressiveness.

Number (6), while it does apply to most domesticated species, also has exceptions, most notably in the domestic cat and ferret, which are both descended from strictly solitary wild ancestors but which tolerate and even seek out social interaction in their domestic forms. Feral domestic cats, for example, naturally form colonies around concentrated food sources and will even share prey and rear kittens communally, while wildcats remain solitary even in the presence of such food sources. Zoologist Marston Bates devoted a chapter on domestication in his 1960 book The Forest and the Sea, in which he talks a great deal about how domestication alters a species: Dispersal mechanisms tend to disappear for the reason stated above and also because people provide transportation for them. Chickens have practically lost their ability to fly. Similarly, domestic animals cease to have a definite mating season, and so the need to be territorial when mating loses its value; and if some of the males in a herd are castrated, the problem is reduced even further. What he says suggests that the process of domestication can itself make a creature domestic able. Besides, the first steps towards agriculture may have involved hunters keeping young animals, who are always more impressionable than the adults, after killing their mothers.

Another strong factor deciding whether a species will be considered for domestication is quite simply the availability of more suitable (or even better already domesticated) alternatives. For example a community that had been introduced to domestication by neighboring peoples will generally find it much more practical, economical and time saving to import already domesticated species than experiment with wild animals (even if they are of the same species). Generally speaking, the species of animals originally domesticated by early humans in the interconnected landmasses of Eurasia and Africa were far superior, both in working capacity and in food production, to the species found in the other continents, namely the Americas and Oceania.

The boundaries between surviving wild populations and domestic clades can be vague. A classification system that can help solve this confusion surrounding animal populations might be set up on a spectrum of increasing domestication.

Wild

These populations experience their full life cycles without deliberate human intervention.

Rose in Captivity/Captured from Wild (in zoos, botanical gardens, or for human gain): These populations are nurtured by humans but (except in zoos) not normally bred under human control. They remain as a group essentially indistinguishable in appearance or behavior from their wild counterparts. Examples include Asian elephants, animals such as sloth bears and cobras used by showmen in India, and animals such as Asian black bears (farmed for their bile), and zoo animals, kept in captivity as examples of their species. (It should be noted that zoos and botanical gardens sometimes exhibit domesticated or feral animals and plants such as camels, mustangs, and some orchids).

Raised Commercially (Captive or Semi-Domesticated)

These populations are ranched or farmed in large numbers for food, commodities, or the pet trade, commonly breed in captivity, but as a group are not substantially altered in appearance or behavior from their wild cousins. Examples include the ostrich, various deer, alligator, cricket, pearloyster, raptors used in falconry and ball python. (These species are sometimes referred to as partially domesticate.)

Domesticated

This classification system does not account for several complicating factors: genetically modified organisms, feral populations, and hybridization. Many species that are farmed or ranched are now being genetically modified. This creates a unique category because it alters the organisms as a group but in ways unlike traditional domestication. Feral organisms are members of a population that was once raised under human control, but is now living and multiplying outside of human control. Examples include mustangs. Hybrids can be wild, domesticated, or both: a liger is a hybrid of two wild animals, a mule is a hybrid of two domesticated animals, and a beefalo is a cross between a wild and a domestic animal.

A great difference exists between a tame animal and a domesticated animal. The term "domesticated" refers to an entire species or variety while the term "tame" can refer to just one individual within a species or variety. Humans have tamed many thousands of animals that have never been truly domesticated. These include the elephant, giraffes, and bears. There is debate over whether some species have been domesticated or just tamed. Some state that the elephant has been domesticated, while others argue the cat has never been. Dividing lines include whether a specimen born to wild parents would differ in appearance or behavior from one born to domesticated parents. For instance a dog is certainly domesticated because even a wolf (genetically share a common ancestor with all dogs) raised from a pup would be very different from a dog, in both appearance and behavior. Similar problems of definition arise when domesticated cats go feral. Since the process of domestication inherently takes many generations over a long period of time, and the spread of breed and husbandry techniques is also slow, it is not meaningful to give a single "date of domestication". However, it is believed that the first attempt at domestication of both animals and plants were made in the Old World by peoples of the Mesolithic period. The tribes that took part in hunting and gathering wild edible plants, started to make attempts to domesticate dogs, goats, and possibly sheep, which was as early as 9000 BC. However, it was not until the Neolithic Period that primitive agriculture appeared as a form of social activity, and domestication was well under way. The great majority of domesticated animals and plants that still serve humans were selected and developed during the Neolithic Period; a few other examples appeared later. The rabbit for example, was not domesticated until the middle Ages, while the sugar beet came under cultivation as a sugar-yielding agricultural plant in the 19th century. As recently as the 20th century, mint became an object of agricultural production, and animal breeding programs to produce high-quality fur were started in the same time period.

The methods available to estimate domestication dates introduce further uncertainty, especially when domestication has occurred in the distant past. So the dates given here should be treated with caution; in some cases evidence is scanty and future discoveries may alter the dating significantly.

Dates and places of domestication are mainly estimated by archaeological methods, more precisely archaeozoology. These methods consist of excavating or studying the results of excavation in human prehistorically occupation sites. Animal remains are dated with archaeological methods, the species they belong to is determined, the age at death is also estimated, and if possible the form they had, that is to say a possible domestic form. Various other clues are taken advantage of, such as slaughter or cutting marks. The aim is to determine if they are game or raised animal, and more globally the nature of their relationship with humans. For example the skeleton of a cat found buried close to humans is a clue that it may have been a pet cat. The age structure of animal remains can also be a clue of husbandry, in which animals were killed at the optimal age.

New technologies and especially mitochondrial DNA which is simple DNA found in the mitochondria that determine its function in the cell provide an alternative angle of investigation, and make it possible to re-estimate the dates of domestication based on research into the genealogical tree of modern domestic animals.

It is admitted for several species that domestication occurred in several places distinctly. For example, research on mitochondrial DNA of the modern cattle *Bos Taurus* supports the archaeological assertions of separate domestication events in Asia and Africa. This research also shows that Bos Taurus and Bos indicus haplotypes are all descendants of the extinct wild ox Bos primigenius. However, this does not rule out later crossing inside a species; therefore it appears useless to look for a separate wild ancestor for each domestic breed.

The first animal to be domesticated appears to have been the dog, in the Upper Paleolithic era. This preceded the domestication of other species by several millennia. In the Neolithic a number of important species such as the goat, sheep, pig and cow were domesticated, as part of the spread of farming with characterises this period. The goat, sheep and pig in particular were domesticated independently in the Levant and Asia. There is early evidence of beekeeping, in the form of rock paintings, dating to 13,000 BC.

Recent archaeological evidence from Cyprus indicates

domestication of a type of cat by perhaps 9500 BC. The earliest secure evidence of horse domestication, bit wear on horse molars at Dereivka in Ukraine, dates to around 4000 BC. The unequivocal date of domestication and use as a means of transport is at the Sintashta chariot burials in the southern Urals, c. 2000 BC. Local equivalents and smaller species were domesticated from the 26th century BC.

The availability of both domesticated vegetable and animal species increased suddenly following the voyages of Christopher Columbus and the contact between the Eastern and Western Hemispheres. This is part of what is referred to as the Columbian Exchange.

Approximate dates and locations of original domestication

Species	Date	Location
Dog (Canis lupus familiaris)	Prior to 33000 Bp	Eurasia
Sheep (Ovis orientalis aries)	between 11000 BC and 9000 BC	Southwest Asia
Pig (Sus scrofa domestica)	9000 BC	Near East, China, Germany
Goat (Capra aegagrus hircus)	8000 BC	Iran
Cow (Bos primigenius taurus)	8000 BC	India, Middle East, and North Africa
Cat (Felis catus)	7500 BC	Cyprus and Near East
Chicken (Gallus gallus domesticus)	6000 BC	India and Southeast Asia
Guinea pig (Cavia porcellus)	5000 BC	Peru
Donkey (Equus africanus asinus)	5000 BC	Egypt
Domesticated duck (Anas platyrhynchos domesticus)	4000 BC	China
Water buffalo (Bubalus bubalis)	4000 BC	India, China
Horse (Equus ferus caballus)	4000 BC	Eurasian Steppes
Dromedary (Camelus dromedarius)	4000 BC	Arabia
Llama (Lama glama)	6000 BC	Peru
Silkworm (Bombyx mori)	3000 BC	China
Reindeer (Rangifer tarandus)	3000 B	Russia
Rock pigeon (Columba livia)	3000 BC	Mediterranean Basin
Goose (Anser anser domesticus)	3000 BC	Egypt
Bactrian camel (Camelus bactrianus)	2500 BC	Central Asia

Contd...

Species	Date	Location
Yak (Bos grunniens)	2500 BC	Tibet
Banteng (Bos javanicus)	Unknown	Southeast Asia
Gayal (Bos gaurus frontalis)	Unknown	Southeast Asia
Alpaca (Vicugna pacos)	1500 BC	Peru
Ferret (Mustela putorius furo)	1500 BC-	Europe
Muscovy Duck (Cairina momelanotus)	Unknown	South America
Guineafowl	Unknown	Africa
Common carp (Cyprinus carpio)	Unknown	East Asia
Domesticated turkey (Meleagris gallopavo)	500 BC	Mexico
Goldfish (Carassius auratus auratus)	Unknown	China
European Rabbit (Oryctolagus cuniculus)	AD 600	Europe

Second Circle :

Species	Date	Location
Zebu (Bos primigenius indicus)	8000 BC	India
Honey bee	4000 BC	Multiple places
Asian Elephant (Elephas maximus) (endangered)	2000 BC	Indus Valley civilization
Fallow Deer (Dama dama)	1000 BC	Mediterranean Basin
Indian Peafowl (Pavo cristatus)	500 BC	India
Barbary Dove (Streptopelia risoria)	500 BC	North Africa
Japanese Quail (Coturnix japonica)	1100-1900	Japan
Mandarin Duck (Aix galericulata)	Unknown	China
Mute Swan (Cygnus olor)	1000-1500	Europe
Canary (Serinus canaria domestica)	1600	Canary Islands, Europe

Domestication of Livestock Species

Modern instances :

Species	Date	Location
Fancy rat (Rattus norvegicus)	1800s	UK
Fox (Vulpes vulpes)	1800s	Europe
European Mink (Mustela lutreola)	1800s	Europe
Budgerigar (Melopsittacus undulatus)	1850s	Europe
Cockatiel (Nymphicus hollandicus)	1870s	Europe
Zebra Finch (Taeniopygia guttata)	1900s	Australia
Hamster (Mesocricetus auratus)	1930s	United States
Silver Fox	1950s	Soviet Union
Muskox (Ovibos moschatus)	1960s	United States
Corn Snake (Pantherophis guttatus guttatus)	1960s	United States
Ball python (Python regius)	1960s	Africa
Madagascar hissing cockroach (Gromphadorhina portentosa)	1960s	Madagascar
Red Deer (Cervus elaphus)	1970s	New Zealand
Hedgehog (Atelerix albiventris)	1980s	United States
Sugar Glider (Petaurus breviceps)	1980s	Australia
Skunk (Mephitis mephitis)	1980s	United States

Chapter - 3

Classification of Livestock Breeds

Classification of Breeds

Phylum	Cordata
Sub-phylum	Craniata (Vertebrata)
Class	Mammalia
Sub-class	Theria (Viviparous)
Infra-class	Eutheria (Placenta)
Order	Ungulata (hoofed mammals)
Sub-order	Artiodactyles (even-toed)
Sub-division	Pecora (true ruminants)
Family	Bovidae (hollow -horned)
Genus	Bos
Species	
taurine group	taurus (European cattle – without hump) indicus (Indian cattle-humped)
bibovine group	gaurns (gaur), frontalis (gayal), sondaians (banteng)
bisotine group	grunniens (yak), bonasians (European bison), bison (American bison)
bubaline group	caffer (African buffalo), bubalis (Indian reverine buffalo), mindorensis (Mindora buffalo), depressicornis (Celebes buffalo)

The Bos Taurus is again Divided into Three Subgroups

Bos primigenins: Strong horns, narrow fore head. Example- Angus, Ayrshire, Short-horn, Holstein Friesian, Red Poll.

Bos longifrons: Broad and dished fore head. Example - Jersey, Guernsey, Brown Swiss.

Bos brachycephalus: Short and broad head. Example - Canadian, Hereford, Kerry.

Definition of Breed

A breed is a group of animal's related similar characters like general appearance, size, features and configuration etc. A group of individual which have certain common characteristicsthat distinguish them from other groups of individuals is known as Species.

Types of Breeds

Domestic cattle belong to the family Bovidae, sub-family Bovinae and can be classified into Bos Taurus and Bos indicus. They have 30 pairs of chromosomes, interbreed and are distributed throughout the tropics. The domestication of Bos-taurus cattle took place some 8000to 9000 years ago and Homitic longhorn and shorthorn types are believed to be their ancestors.

The origin of Bos indicus breeds (humped cattle) were in western Asia. Both humped and hump less cattle were introduced to Africa from western Asia and into America and Australia from Europe by the immigrants.The European cattle Bos-taurus were introduced in the tropics to be raised as pure-bred and crossbred with indigenous breeds. As a result of crossing of native cattle with European dairy breeds, large numbers of crossbreeds have been produced in various tropical countries, which are being used in selection programs. Zebu Bos indicus cattle were introduced into United States in the nineteenth century for crossbreeding with European breeds. Breeds resulting from crosses are used in the southern regions of North America and tropical South America. Most of the cattle breed in the tropics evolved, through natural selection, for adaptability and survival to local environments. Often, breeds resemble each other with slight morphological differences, but because of constant inbreeding in one locality, independent breeds have evolved. In general, the cattle from drier regions are well built and those from heavy rainfall areas, coastal and hilly regions are of smaller build.

Classification of Cattle Breeds

Most indigenous cattle breeds in the tropics are multipurpose (milk, meat, draught) and that only a few breeds have good milk potential. Physical and economic parameters for some of the important indigenous dairy breeds and new crossbred types developed in the tropics are discussed. Indian cattle breeds of cattle are classified in to three types as under: a) Milch breeds: b) Dual Purpose breeds: c) Draught breeds

Milch Breeds

The cows of these breeds are high milk yielder and the male animals are slow or poor worker. The examples of Indian milch breeds are Sahiwal, Red Sindhi, Gir, Tharparker and Rathi. The milk production of milk breeds is on the average more than 1600 kg. per lactation.

Dual Purpose Breeds

The cows in these breeds are average milk yielder and male animals are very useful for work. Their milk production per lactation is 500 kg to 1500 kg. The examples of this group are Ongole, Hariana, Kankrej, Krishna valley, Goalo, Deoni and Mewati.

Draught Breeds

The male animals are good for work and cows are poor milk yielders and their milk yield on an average is less than 500 kg per lactation. They are usually white in colour.

Milch Breeds

Sahiwal

This is one of the best indigenous breeds of dairy cattle. The original tract of this breed is Montgomery district in Pakistan, but animals of this breed are also found in Ferozepur, Amritsar, Gurdaspur districts of Punjab. Several breeding herds are maintained in Punjab, Delhi, Haryana, Uttar Pradesh, Chhattisgarh and Madhya Pradesh. The cows are red and light brown in colour, but some animals with white patches are also found. The milk yield under village condition is 1350 kg. Whereas under commercial dairy farms is 2100 kg. The average fat and SNF content of the milk is around 5.0 and 9.2 %, respectively. Likewise, age at first calving is between 32 to 36months and the calving interval is about 15 months. It is a medium sized breed, symmetrical body, broad fore head, thick short horns and fine loose skin Dewlap is fine and ample in the male. Chest is broad and deep, legs proportionate to size with good feet. In the male the sheath is pendulous, the tail is long with a black switch, and udder is large, broad and fine skin with prominent veins. Teats are good, uniform in size squarely placed. Milk veins are large and prominent. Sahiwal animals have been exported to Sri Lanka, Kenya, and the West Indies and many other Latin American countries. A new breed called Jamaica Hope has been evolved using Sahiwal x Jersey crossbreeds

Gir

Although, the breed is native of Gujarat, it is also found in Maharashtra and Rajasthan States in India. The peculiar features of the breed are a protruding-broad and long forehead, and pendulous forward turned ears. The popular colour is white with dark red or chocolate-brown patches distributed all over the body. Entire red animals are also encountered although it is usually mottled with yellowish-red to almost black patches. The animals are medium sized with proportionate body. The head is moderately long, massive and the forehead bulging. The face is narrow and clean. The muzzle is square and black. The eyes are placed higher up in line with root of ears. Ears are large and pendulous. The horns are black, medium sized, shapely round medium heard, well set apart, and peculiarly curved. They take a down ward and backward curve and inline a little upwards and forward taking a spiral inward sweep, finally ending in a fine taper. Dewlap is thin and hanging not pendulous. Chest is deep, full and well developed. Legs are well proportionate and muscular. The hump is medium sized. The barrel is deep, long and proportionate. The back is long, strong and wide. The tail is long touching the ground. The udder is of medium size. Average weight of the male and female is 545 kg. and 386 kg., respectively. Gir cows are good milkers and milk yield ranges from 1200 to 1800 kg per lactation. The age at first calving varies from 45 to 54 months and the inter-calving period from 515 to 600 days. The Gir breed has been exported to other parts of the world. In Brazil where large herds are found, it is known as Gyr. Brazil has also evolved a strain called Indubrasil which is a cross between Gir and Kankrej. Gir animals are highly prized by Brazilian breeders. Gir animals have also been exported to the USA, especially Texas, Florida and Lousiana states.

Redsindhi

The Red Sindhi cattle have somewhat similarity in breed characteristics to that of Sahiwal, but are smaller in size with compact body frame. This breed is from Sindh in Pakistan. The colour of the breed is deep dark red. The bulls are much darker than cows. A white marking on the forehead is common. The animals are medium sized, compact and symmetrical. The head is of moderate size, forehead is broad and poll is prominent in between horns. Face is medium in length with well developed square black muzzle. Eyes are fairly large, and clear. Ears are medium sized, fine and alert. The horns are short and thick. Dewlap is abundant in both males and females and hangs in folds, chest is broad and deep. Legs are medium in size. Tail is slender with black switch. The udder is large size with medium sized teats. Milk veins are well developed. In India pedigree herds are found in Mysore, Tamilnadu Uttarakhand and Orissa. Average weight of the male is 420 kg and that of the female is 341 kg. The milk yield in the institutional herds ranges from 1250 to 1800 kg per lactation. Age at first calving is 39 to 50 months and the calving interval 425 to 540 days. Red Sindhi animals have also been exported to many other parts of the world including Sri Lanka, Tanzania, the Philippines, the United States, Malaysia, Iraq, and Burmaand Indo-China. Red Sindhi females have been used in crossbreeding with Brown-Swiss and Jersey to develop new breeds such as Karan Swiss and Jer Sind in India.

Tharparker

It is an important cattle breed raised primarily for its milking potential .The original habitat of this breed is Tharparkar district in the Province of Sind, Pakistan. The breed is also found in the adjoining tracts in Rajasthan State in India, particularly around Jodhpur and Jaisalmer where excellent milch specimens are found. This is a medium-sized compact breed. The males are also good draught animals. The milk yield in cows ranges from 1800 to 2600 kg per lactation, age at first calving is from 38 to 42 months and the inter calving period is from 430to 460 days. Lactation length is 285 days and dry period 140 days. Milk fat is about 4.88 % and SNF 9.2 %.

Rathi

Rathi cattle are named after a pastoral tribe of Rajasthan called Raths who lead a nomadiclife. The home tract of this breed lies in the heart of Thar Desert. Rathi breed is a mixture of Sahiwal, Red Sindhi, Tharparkar and Dhanni breeds with a preponderance of Sahiwal blood. The animals are of medium size with a symmetrical body and a short and smooth body coat. The animals have brown colour with white patches and some animals with complete brown or black coat colour with white patches are also found. Horns are short to medium curving outwards, upwards and inwards, ears are of medium size, voluminous dewlap and large naval flap. Their udder and teat are well developed with a prominent milk vein. The females are docile and good milkers (1325 to 2093 kg per lactation). Calving interval ranges between 445 and 617 days. The average age at first calving ranges from 36 to 52 months and inter calving period ranges from 450 to 620 days.

Dual Purpose Breeds

Ongole

The home of this breed is Ongole tract and Nellore districts of Andhra Pradesh. The cows are good for milk production usually docile and bullocks are very powerful and good for heavy plough and Cart work. The colour of the animal is white. The forehead is broad and prominent between eyes, black kazal marking around the eyes are common. The horns are short and stumpy. Loose horns are common in this breed. The tail is long with black switch reaching below the hocks. The udder is broad, extends well forwards and high up with moderate an even sized quarters and teats are average size mammary veins are prominent. The bullocks are very powerful and good. Ongole is one of the heaviest breeds. The weight of the male varies from 545kg to 682 kg. and that of the female is 432 to 455 kg. The average milk yield is 1600 kg in lactation.

This is essentially a large muscular breed suitable for heavy draft work. The age at first calving is 38 to 45 months and the intercalving period 470 to 530 days. Excellent specimens of this breed have been exported to Brazil where large herds now exist. They are known as Nellore in Brazil. This breed has also been exported to Sri Lanka, Fiji, Indo-China, Indonesia, Malaysia and the United States. The famous Santa Gertrudis breed evolved in Texas, USA includes Ongole blood.

Hariana

The native breeding tract of this breed encompasses large parts of Rohtak, Hisar and Gurgaon district of Haryana State and is a prominent dual-purpose breed of north India. It has been extensively used in grading up of non-descript cattle particularly to improve their draught capacity in the Indo-Gangetic plains. The colour of the breed is white or light grey. Hariana cattle are characterised by a long and narrow face, flat forehead and well marked bony prominence at the centre of the poll. The muzzle is usually black. This eyes are large and bright expressive but not prominent in mature bulls. The horns are short and fine or moderately long, and they are generally 4 t0 9 inches, long thinner in females than in males. Dewlap is small without flashy folds and large in males. The chest is well developed. The udder is capacious and extends well forward with a well-developed milk vein. The teats are well developed, proportionate and medium sized. Good specimens of cows yield up to 1500 kg of milk per lactation. The age at first calving is 40 to 60 months depending on management and feeding conditions. The intercalving period varies from 480 to 630 days. The hump is large in males and medium sized in females. Legs are moderately long and lean and feet are small, hard and well shaped. In the males the sheath is short and tight and in the females the navel flap is not prominent. Tail is short, thin, reaching below the hock and tapering with black switch. Teats are medium sized and proportionate. The average weight of males is 371 to 490 kg and that of the females is 265 kg. The average milk yield of cows is 909 to 1364kg. The bullocks are good for ploughing and road transport.

Kankrej

The home of this breed is Gujarat. It is one of the heaviest breed in India. The colour of the female is Silver gray, iron or black. The males are darker than the females. The forehead is broad slightly dished in the center. The horns are thick, strong and curved. The base of the horns is covered with skin. The body is powerful, with broad chest. Straight back, well developed hump, pendulous sheath in males, and tail is of moderate length with black switch extending below the hock. Dewlap is thin and pendulous and hump is large and prominent. In cow's udder is well shaped and slightly developed and carried more forward than behind. The average weight of the male is 455 to 682 kg. and of the female is 409 to 455 kg. The average milk yield is 1750kg. and milk fat is around 4.8 %. The lactation length averages 295 day sand the calving interval is around 490 days. The age at first calving is 36 to 42 month .This breed has been exported to a number of countries to be raised as pure-bred and for crossbreeding. Today excellent herds of this breed are found in Brazil where it is known as Guzerat. Many beef breeds in some Latin American countries and southern states of the USA have some inheritance of Kankrej.

Deoni

This is a very popular breed of cattle of Marathwada region in Maharashtra and adjoining parts of Karnataka and Andhra Pradesh states. The body colour of the animal is white and black patches or red and white patches. The animals resemble Gir breed to a great extent. The forehead is less prominent. The ears are long and pendulous. The chest is heavy and deep, the dewlap is well developed and in the males the sheath is pendulous. The head is medium sized, prominent forehead, the horns curving outwards and backwards. A wedge shaped barrel and well placed. This breed is considered to have admixture of Gir, Dangi and local cattle. The horns emerge from the side of the poll in outward and upwards direction, slightly backward and again curving upward. Deoni animals are fairly good milk producers and the milk yield is 700 kg. under village condition and 1000 kg. in well developed commercial farms. Milk contains 4.3 % fat, 9.7% SNF and 14 % total solids. The age at first calving is 46 months. Calving interval averages 450 days. The animals are docile and calm. The bullocks are large sized and good for heavy work.

Draught Breeds

Draught breeds are four types

- *Short horned:* White or Grey with long coffin shaped skull. Example Nagori, Bachur
- *Lyre horned:* Grey with wide fore head. Example Malvi, Kerigarh
- Small black, red or dum cattle with large patches of white marking found in himalayan region. Example Ponwar, Siri
- *Mysore type:* Prominent fore head with long and pointed horns which rise closs together. Example Hallikar, Umbalachery, Alambadi, Pulikulam, Amritmahal, Burgur, Khillari , Kangayam.

Malvi

The breed is found in Malwa tract in Madhya Pradesh and Rajasthan. The bullocks are known for their draft qualities and the cows are poor milkers. The colour of the animals is white to light grey, with black markings on neck, shoulders, hump and quarters. The colour changes with age. The head is small and the face dished. The body is deep, short arid compact with short legs and the tail touching the fetlocks. Ears are short and alert. The sheath in the male and navel flap in the female are short. The horns are massively built. Black, upright and pointed at tips.

Hallikar

The home tract of this breed is Karnataka but the breed is widely distributed in South India. The colour of the animal is dark or light grey with white patches round the face and dewlap. The bullocks are suitable for both for road and field work. The head is long with bulging forehead furrowed in the middle. Horns are close together and spring perpendicularly from the head, carried backward with a graceful sweep on each side of the neck and curving upwards and terminate in sharp point. The body is long and compact with long and slender legs. The novel flap is tucked up and tail is thin.

Amrit Mahal

Amrit Mahal literally means the department of milk. Originally the rulers of Mysore State had started an establishment of cattle collected from the prevalent types of cattle within the area for the supply of milk and milk products to the palace. The home of this breed is Karnataka. The cattle of Amrit Mahal establishment originally comprised three distinct varieties: Hallikar, Hagalvadi and Chitradoorg. Prior to 1860, it seems that these three varieties were maintained separate from each other. In 1860, the whole establishment was liquidated for reasons of economy. By the year 1866, it was realized that an establishment for the supply of cattle was necessity, and, during the year, a herd was again established. Thus, the foundation cattle from which the Amrit Mahal breed was developed were of the Hallikar and closely related types.

The coloring of Amrit Mahal cattle is usually some shade of gray varying from almost white to nearly black, and in some cases white-gray markings of a definite pattern are present on the face and dewlap. The muzzle, feet and tail switch are usually black, but in older animals the color looks lighter. This is the best breed in India for drought purpose. The bullocks are suited for quick transport and the cows are poor milkers. The animals are active and fine in temperament. The barrel is long and well rounded, and the novel flap is tacked up. The head is well shaped, narrow, and the fore head is deeply furrowed. The eyes are bright. The legs are well proportioned and medium in length. The hooves are hard black with narrow clefts. The tail is fine and moderate in length. The udder is small compact with small hard teats. At the same time; the bullocks were utilized for the movement of army equipage. The bullocks were regularly classified as gun bullocks, pack bullocks, plow bullocks, etc.

Kangayam

The home tract of this breed is Tamilnadu. The colour of the animals is white and grey but the cows are white with black markings in front of fetlocks or on knees. The bulls are good for hard work and the cows are poor milkers. This is a medium sized draft breed. The bullocks are strong active and suited for heavy work and road transport. The head is short with a broad forehead. Horns curving outwards backwards and complete a circle at the point. The legs are short; the sheath in the male is small and the moderately long. The udder is medium sized. Calves are red at birth and the colour changes to white when they are about 4 months of age.

Gaolao

The Gaolao cattle fit into the group of short horned, white or light-gray in color, with a long coffin-shaped skull, orbital arches not prominent and with a face slightly convex in profile.

It is also observed that the native home of the breed is located along the route taken by the Rig Vedic Aryans from the Northern passes through Central India to the South

There is a close similarity between the Ongole and the Gaolao except the latter are much lighter, with greater agility. It is said that the Marathas developed this breed into a fast-trotting type suitable for quick army transport in the hilly areas of Gondwana, Madhya Pradesh. It was used mainly for military purposes by the Maratha army when invading the local Gond Kingdom. Old historical records show that the breed had fair milk-producing capacity, but during the last two centuries selection has been directed mainly towards developing a capacity for quick draft. The breed is found principally in the districts of Wardha, Nagpur and Chindwara.

Gaolao animals are of medium height, or rather light build and tend to be narrow and long. The head is markedly long and narrow with a straight profile usually tapering towards the muzzle and somewhat broader at the base of the horns. The forehead is usually flat, though it appears to recede at the top, giving a slightly convex appearance. The eyes are almond-shaped and placed slightly at angles.

The ears are of medium size and are carried high. The horns are short and stumpy, blunt at the points and commonly slope slightly backwards. The neck is short, with a moderately well-developed hump, which is usually loose and hangs on one side. The hind quarters are slightly drooping. Limbs are straight and muscular. Hooves are of medium size, hard and durable, and suited to hard road and hillside work. The dewlap is large but the sheath is only moderately developed. The skin is thin but loose. The tail is comparatively short, reaching only a little below the hocks. Females are usually white and males gray over the neck, hump and quarters.

Kankrej

The Kankrej breed of cattle gets its name from a territory of that name in North Gujarat of Bombay Province, India. It also Known as: Bannai, Nagar, Talabda, Vaghiyar, Wagad, Waged, Vadhiyar, Wadhiar, Wadhir, Wadial. The breed comes from southeast of the Desert of Cutch in western India, particularly along the banks of the rivers Banas and Saraswati which flow from east to west and drain into the desert of Cutch. In Radhanpur State, which is adjacent to the Kankrej tract, the breed is known as Wadhiar. In Cutch State it is known as Wagad or Wagadia, taking its name from the community of herdsmen who breed these cattle.

The Kankrej is one of the heaviest of the Indian breeds of cattle. Color varies from silver to gray to iron gray or steel black. Newly born calves have rust red-colored polls, this color disappearing within 6 to 9 months. Fore quarters, hump and hindquarters are darker than the barrel, especially in males. The switch of the tail is black in color. The forehead is broad and slightly dished in the center. The face is short, and the nose looks slightly upturned. The strong lyre-shaped horns are covered with skin to a higher point than in other breeds. The ears are very characteristic, being large, pendulous and open. The legs are articularly shapely and well-balanced and the feet small, round and durable. They are active and strong. The hump in the males is well-developed and not so firm as in some breeds. The dewlap is thin but pendulous and males have pendulous sheaths. Pigmentation of the skin is dark and the skin is slightly loose and of medium thickness. Hairs are soft ad short. The Kankrej cattle are very highly prized as fast, powerful draft cattle. They are also fair producers of milk. These cattle are resistant to Tick fever and they show very little incidence of contagious abortion and tuberculosis. It has also been observed that the red color is recessive.

Nagori

Nagori cattle are prevalent in the former Johrpur State, now a part of the State of Rajasthan in India. Nagori cattle are classified into the short-horned white or light gray cattle with a long coffin-shaped skull, orbital arches which do not prominent, and their face is slightly convex in profile.

It has been suggested that probably the blood of gray lyre-horned cattle might have entered into the composition of Nagori cattle. Taking into consideration the proximity of the native homes of the Hariana in the north and northeast and Kankrej in the south and southwest, it seems reasonable to suppose that Nagori cattle may have evolved from these two groups.

Frequency of famines in its native home has necessitated extensive movements of the cattle to other regions in search of fodder, and this has no doubt led to frequent intermixture.

Generally the Nagori cattle are fine, big, upstanding, active and docile, with white and gray color. They have long, deep and powerful frames, with straight backs and well-developed quarters.

There is throughout the Nagori breed a tendency to legginess and lightness of bone, though the feet are strong. It is supposed that this characteristic has given the breed its agility and ease of movement. The face is long and narrow but the forehead is flat and not so prominent. The eyelids are rather heavy and overhanging and the eyes are small, clear and bright. The ears are large and pendulous. The horns are moderate in size and emerge from the outer angles of the poll in an outward direction and are carried upwards with a gentle curve to turn in at the points.

The neck is short and fine, and looks powerful. The dewlap is small and fine. The hump in the bulls is well-developed but not so firm and thus in many cases hang over. The shoulders and forearms look muscular and powerful. The legs are straight with hooves compact, strong and small. The tail is of moderate length reaching just below the hocks and terminating in a tuft of black hair. The sheath is small. The skin is fine and slightly loose. The cows usually have well-developed udders with large teats. The Nagori breed is one of the most famous trotting draft breeds of India and is generally appreciated for fast road work. As such, more attention has been paid by the breeders towards producing an agile yet powerful animal with a great deal of endurance.

Nagori cattle are famous as trotters, being used all over Rajasthan in light iron-wheeled carts for quick transportation. They are also worked for all agricultural purposes, such as plowing, cultivation drawing water from wells and transportation of field produce to markets.

Nelore

The most distinctive characteristic of this breed is the presence of a prominent "hump" behind their neck, but there are many other fundamental differences between the Nelore and the European breeds.

There has never existed in India a breed called Nelore. This name corresponds to a district of the old Presidency of Madrás, now belonging to the new State of Andhra Pradesh. It was in Brazil that some authors started to use the name Nellore as a synonym to Ongole, the Indian breed that contributed most to the creation of the Nelore. The history of the Ongole dates back 2,000 years before Christian times. It was the Aryan people that brought the ancestors of the Nelore to India, where they were submitted to extreme weather conditions. The arid lands of Beluchistan, the cold winters of Punjab, the alluvial lands of Ganges and the torrid lands by the Bengal Sea provided the Ongole breed with the adaptation genes that are now favorably expressed in the modern Nelore.

Brazil has become the largest breeder of Nelore, and from there the breed was exported to Argentina, Paraguay, Venezuela, Central America, Mexico, United States and many other countries. In all those places, the contribution of the Nelore was remarkable, whether through purebred selection within the breed or through crosses with local breeds, many times of European origin. The creation of the Nelore Herd Book and the definition of the breed standards in Uberaba, 1938, were of great relevance in the formation of the Nelore. In 1960, 20 animals were imported, and in 1962, the last and most relevant purchases of live animals from India authorized by the Brazilian Government, 84 Ongoles were imported. These became founders of important breeding lines like Godahvari, Karvadi and Taj Mahal, and were decisive to the great expansion of the Brazilian herd in the last 30 years, going from 56 million in 1965 to 160 million in 1995, 100 million of which are Nelore.

It is said that there is no ideal breed, and that every breed has strong points and none is better for all important economic traits, the Nelore is certainly the best alternative for economic beef production in the tropics, which are responsible for 65% of the world's bovine population.

The main advantage that the Nelore has over other breeds of beef cattle is itshardiness. Calves are alert, with an active behavior, standing up and suckling soon after they are born, without any need for constant human intervention. The Nelore has notable physical strength and is unexcelled in its ability to thrive under harsh climatic, nutritional and sanitary conditions, frequent in the tropics. Because of their hardiness and rustling ability, Nelores surpass all other breeds under conditions of poor range and drought.

The Nelore has a loose skin with sweat glands that are twice as big and 30% more numerous than those of the European breeds. The Nelore's black skin, covered by a white or light gray coat helps filtering and reflecting harmful sun rays. Its low level of metabolism also contributes to heat resistance, as the Nelore feeds less but often, generating less internal heat. Nelores possess natural resistance to various insects, as its skin has a dense texture, making it difficult for blood sucking insects to penetrate. Nelores also have a well developed subcutaneous muscle layer which enables them to remove insects simply by shaking their coat.

Nelores have long, deep bodies with clear underlines, keeping vulnerable parts out of the way of infection. Cows have small udders and short teats, while bulls' sheaths are also short. These characteristics contribute to the breed's reproductive efficiency. Nelore dams have a long and prolific reproductive life, pronounced mothering ability, and plenty of milk for their calves. The Nelore cows calve very easily due to their greater frame, wide pelvic opening and larger birth canal, which reduce the incidence of distocia.

Nelore dams have highly developed maternal instinct throughout the whole milking period, which is of great importance for extensive breeding systems. They lick their newborn, put them to suckle and look for a safe place to hide them from predators. The active and vivid disposition of the Nelore is largely responsible for their unusual thriftiness, hardiness and adaptability to a wide range of feed and climate. Nelores like affection and quickly respond to kind handling methods, becoming extremely docile.

Nimari

Nimari cattle show a mixture of Gir and Khillari (Tapi Valley strain) breeds. The breed has taken the coloration from the Gir as well as its massiveness of frame and the convexity of the forehead.

It has acquired the hardiness, agility and temper of the Khillari with the formation of feet and occasional carroty color of the muzzle and hooves. Starting from Barwani and Khargone districts of Madhyabarat, the breed spreads into Khandwa, and parts of Harda of Madhya Pradesh. It is also bred in adjacent parts of Bombay State. In the Satpura ranges of Madhya Pradesh there is a strain of cattle known as Khamla, which is much smaller in size but very akin to the Nimari. In addition, the Khamgaon strain found in Berar may be an offshoot of the Nimari. This breed of cattle is prized for draft work, though few animals show evidence of fair milking qualities.

The animals are well-proportioned and compact in appearance. In general they are red in color with large splashes of white on various parts of the body. In the Khamgaon strain the color is occasionally black or light red and white. In the Khamla strain it is red with a violet tinge and white or yellow and white.

The head is moderately long with a somewhat bulging forehead; it is carried alertly and gives the animals a graceful appearance. The horns usually emerge in a backward direction from the outer angles of the poll, somewhat in the same manner as in Gir cattle, turning upwards and outwards and finally backwards at the points. Occasionally, the horns are also like the Khillaris in size and shape, with copper color and pointed. The ears are moderately long and wide and are not pendulous. The muzzle in many animals is either copper-colored or amber-colored.

The body is long, with a straight back and moderately arched ribs with the quarters usually drooping to some extent. There is a tendency to prominent hips common to the Gir. The dewlap and sheath are moderately developed, though the sheath is apt to be pendulous.

The hump in bulls is well-developed and apt to be hanging at times. The limbs are straight and clean and the tail is long and thin with a black switch reaching to the ground. Hooves of the animals are strong and can stand rough wear on stony ground. The skin is fine and slightly loose. The cows usually have well-developed udders.

Ongole

The Ongole breed, like other breeds of cattle in India, takes its name from the geographical area in which it is produced. It is also called the Nellore breed for the reason that formerly Ongole Taluk, a division of a district, was included in the Nellore district, but now it is included in the Guntur district. The area is part of the Andhra Pradesh in India.

This breed is included among the gray-white cattle of the north, having white or gray color, stumpy horns and a long coffin-shaped skull. It has a great similarity with the Gaolao breed of Madhya Predesh and also has a resemblance to the Bhagnari type of cattle in the north of India.

This similarity is not surprising in view of the fact that these breeds lie along the path taken by the Rig Vedic Aryans in their march from the north to the south of India.

It is claimed that the finest specimens of the breed are found in the area between the Gundalakama and Alluru rivers in the Ongole and Kandukur taluks, and also in the villages of Karumanchi, Nidamanur, Pondur, Jayavaram, Tungtoor and Karvadi and along the banks of River Musi. They are also famous from the taluks of Vinukonda and Narasraopet.

The Ongoles are large-sized animals with loosely knit frames, large dewlaps which are fleshy and hang in folds extending to the navel flap, and slightly pendulous sheaths. They have long bodies and short necks; limbs are long and muscular.

The forehead is broad between the eyes and slightly prominent. Eyes are elliptical in shape with black eyelashes and a ring of black skin about 1/4 to 1/2 inch wide around the eyes. Ears are moderately long, measuring on an average for 9 to 12 inches, and slightly drooping.

Horns are short and stumpy, growing outwards and backwards, thick at the base and firm without cracks. In some animals the horns are loose; this is probably due to the horn core not growing well.

The hump in the males is well-developed and erect and filled up on both sides and not concave. The skin is of medium thickness, mellow and elastic and often shows black mottled markings.The popular color is white. The male has dark gray markings on the head, neck and hump and sometimes black points on the knees and on the pasterns of both the fore and hind legs. A red or red and white animal of typical conformation is occasionally seen.

They have a white switch of the tail, white eyelashes, a flesh colored muzzle, light colored hooves, dark gray marking on the hindquarters and dark mottle appearance on the body.

Ongole cattle are efficiently used in their native home for both work and milk production. They are usually docile and the bulls are very powerful, suitable for heavy plowing or car work but are not considered to be suitable for fast work or trotting purposes. The cows are fair milkers.

All animals currently used for food and agriculture and the result of Domestication from wild progenitor species like their wild relatives. These Domestic species are continuously evolving albeit at an accelerated rate due to human activity.

Siri

Animals of this breed are found in the hill tracts around Darjeeling (Bengal, India) and in Sikkim and Bhutan. Bhutan is said to be the real home of this breed. It is distributed from that area to the various parts of Sikkim and Darjeeling. The Siri has a hump that is thoracic and muscular-fatty. Presumably Siri cattle have some blood from the cattle in Tibet.

Small cattle with similar black and white markings have been found in Sikong Province of China, which occupies a portion of the Tibetan highlands northeast of Bhutan. Siri cattle crossed with Nepali cattle look like Siri, but they can be distinguished by their color pattern and position of hump and horns. These are known as Kachcha Siri or imitation Siri cattle.

The colours most frequently seen are black and white or extensive solid black, in color patterns similar to that of Holstein-Friesians. The animal carries a thick coat all the year round, and it is generally believed that this protects them from heavy rains and severe cold.

The general form of the animal is massive. The head is small, square cut and well set on. The forehead is wide and flat. The horns are sharp and directed forward and are usually covered with a tuft of long coarse hair. The position of the hump is slightly forward compared with that of other Zebu breeds.

The dewlap is moderately developed and the sheath in the male is tight. Strong legs and feet are characteristics of this breed. The hooves are broad but strong. The udders of the cows are well developed. It is observed that the animals of this breed can stand the rugged conditions of the mountains very well. When the animals are brought down to the plains they do not seem to do so well. Bulls are eagerly sought after for draft purposes due to their size and reputed great strength. They are also used for agricultural work such as plowing, cultivating, threshing, etc.

Mewati

Mewati cattle are found in the tract known as Mewat, but the breed is sometimes called Kosi, due to the large numbers of cattle of this breed sold from the market at Kosi, a small town in the district of Mathura. Mewati cattle are similar in type to Hariana, but show definite evidence of Gir blood. Native habitants of Rath and Nagori cattle being adjacent to Mewat, these two breeds may also have contributed to the formation of the Mewati.

Mewati cattle are usually white in color with neck, shoulders and quarters of a darker shade. Occasionally, individual beasts have Gir coloration. The face is long and narrow with the forehead slightly bulging.Horns emerges from the outer angles of the poll and are inclined to turn backwards at the points. Eyes are prominent and surrounded by a very dark rim.

The muzzle is wide and square and the upper lip thick and overhanging, giving the upper part of the nose a contracted appearance. The muzzle is pitching black in color. The ears are pendulous but not so long. The neck and the whole frame are strong but the limbs are light. The legs are relatively long and the frame of the body gives an impression of being loosely built. The chest is deep but the ribs are flat.The head and neck shows an upright carriage. The dewlap, though hanging, is not very loose. The sheath also is loose but not pendulous. The legs are fine and round with strong, somewhat large hooves, well-rounded in shape. The tail is long, the tuft nearly reaching the heels. Cows usually have well-developed udders. Mewati cattle are in general, sturdy, powerful and docile, and are useful for heavy plowing carting and drawing water for deep wells. The cows are said to be good milkers.

Kherigarh

The Kherigarh cattle are closely tied to the Malvi breed. The breed is mostly found in the Kheri district of Uttar Pradesh, India. Though the horn formation is typical of the lyre-horned Malvi type, the animals of the breed are much lighter in general appearance than the Malvis. The Kherigarh cattle are generally white or gray in color. The face is small and narrow. Horns are thin and upstanding and measure 12 to 18 inches in length in bulls; cows usually have smaller horns. The ears are small and the eyes bright. The neck is short and looks powerful. The hump is well-developed in bulls. The dewlap is thin and pendulous and starts from right under the chin and continues right down to the brisket. The barrel is broad and deep. The sheath is short and moderately tight. Limbs are light. The tail is long, ending in a white switch.

The cattle of this breed are very active and thrive on grazing only. The bullocks are good for light draft and quick, light transport. The cows are poor milkers. It is estimated that these cattle start work when they are about 4 years of age and weigh about 300 kg. It is claimed that a pair of bullocks can haul about 1.5 tons of load in a cart to a distance of 30-35 miles in a day traveling at times 3 to 4 miles per hour.

Umblachery

The Umblachery is found in the region of Thanjanvur, Tamil Nadu in India. It is also Known as: Jathi madu, Mottai madu, Southern Tanjore, Therkuthi madu. It is a draft breed of the zebu type, similar to Kangayam but smaller. They are grey with white points and backlines. Calves are red or brown with white markings. The breed is rare.

The calves of this breed are red or brown in colour at birth. The red colour begins to change to grey at three to four months of age. Total grey colour is generally attained at six to eight months of age. Heifers and cows are grey in colour. In majority of cows dark grey colour is present on face, neck and pelvic regions. Bulls are grey in colour with dark grey on the hump, extremities, fore quarter and hindquarter. They have white star on the face. The switch of the tail is white or partially white in colour. The mean chest girth, body length and height at withers of bulls, bullocks and cows were 145, 118, and 112 cm; 151, 119 and 117 cm and 135, 109 and 105 cm respectively. The principal body measurements reveal that this breed of cattle is of medium size and smaller than other breeds of Tamilnadu.

This breed is suitable for ploughing in marshy paddy fields of the deltaic breeding tract because of its medium size. A pair of bullocks was able to pull 2000-2200 kg over a distance of 20 km in seven hours. Pure breeding of Umblachery breed was done mostly by natural mating.

Vehoor

Known for its high milk yield, fat content and disease resistance, the cow was very popular and abundant in central Kerala till 1960 when the State government took up intensive cross-breeding of native cows with imported exotic bulls for increasing the milk yield.

This was followed by the enactment of the Kerala Livestock Act, 1961; prohibiting maintenance of indigenous bulls, resulting in the Vechoor cattle reaching near extinction by the 1980s. It took another decade for the people to realise and attempt to remedy the situation. Accordingly, the Vechoor Cow Conservation Project was started in 1990 by the University with the assistance of the Indian Council of Agricultural Research (ICAR). At present, a nucleus stock of about 135 cows and bulls having a maximum hump level height of 105 cm and weighing an average 107 kg, are thriving at the University's cattle farm .

The Vechoor cow, credited to having the highest milk yield, calculated in proportion to its body weight and very low feed requirements, is listed as a 'distinct breed' among the 30 breeds of Indian cattle under the ICAR's latest calendar on 'cattle breeds of India' published by the National Bureau of Animal Genetic Resources. Thrissur: The Kerala Agriculture University here has gained international attention by saving the world's smallest cattle of Kerala origin, the "vechoor cow" from near extinction. It has also earned recognition from the Food and Agriculture Organisation (FAO), which has listed the 'Vechoor cow' among the Indian breeds in its domestic animal diversity information system.

Exotic Dairy Cattle Breeds

The European breeds of dairy cattle belong to the species of *Bos Taurus*. They are hump less generally large spread with a fine coat, short ears, without a pendulous dewlap. They are less heat tolerant and less disease resistant as compared to Indian cattle, but are superior in milk production, the exotic breeds of cattle have been used in India on a fairly extensive scale with aview to improve the milk yielding capacity of the indigenous cows. The important European breeds of dairy cattle are Holstein Friesian, Brown Swiss, Jersey, Guernsey and Ayrshire.

Holstein Friesian

The home of this breed is Holland. This is the highest milk yielding breed though the fat percentage of milk is low. This is the best dairy breed among exotic cattle regarding milk yield. On an average it gives 25 Kg. of milk per day, whereas a cross breed H.F. cow gives 10 to15 kg per day. The animals of this breed are the largest among the European breeds. The colour of the animal is clearly defined with black and white markings and switch is always white. The head is long and narrow. The cows are docile. The heifers are bred at 18 to 21 months of age. The calves are stronger, vigorous weighing on an average about

40 kg at birth. Some pure bred animals may be solid black. Holstein Friesian heifers mature much faster than the other European breeds. The ideal body weight of a cow is 682 kg and that of bull is 1000 kg. The cows are heavy milkers and the average lactation yield is 4295 kg with milk fat of 3.4 percent. However, individual animal have yielded 19,995 kg of milk in a lactation period of 365 days. The milk of these animals is used for cheese making as the fat percentage is low.

Brown Swiss

These cattle were developed in mountainous area of Switzerland. The colour of the animal varies from light brown to almost black. The muzzle is of light colour. It is the second-heaviest to the Holstein Friesian breed. Brown Swiss animals have large body size. Heifers mature and reach peak production at an later age than other dairy breeds. The breed was developed for cheese production and so emphasis was given for high milk production with low fat content and the milk fat is 4 %.

Jersey

The home of this breed is Jersey Island. This breed is most popular and widely distributed allover the world. Jersey is the smallest of the European dairy breeds. The heifers are bred at an age14 to 18 months of age. The colour of the animals is brown with variation of brown to black and vary from white spotted to solid in marking. The switch is white or black. The animal is small in size with a good capacity for milk production. The milk fat is high i.e. 5.3% and milk solids are about 15 %. Jersey milk is yellow in colour due to high carotene and is good for butter making. The average milk yield of the cow is 2727 kg. in a lactation However, individual animal have yielded 13.296 Kg. in a year. The age at first calving is 26-30 months and inter calving period is between 13 to 14 months. Daily milk yield is found to be 20 kg. Whereas cross bred Jersey cow gives 8 to 10 Kg per day. In India this breed has acclimatized well especially in the hot and humid areas.

Guernsey

The home of this breed is Guernsey Island. The colour of the animal varies from light brown to almost red with white markings. White markings are usually found on face, legs flank and switch. The nose may be cream or buff coloured having smoky colour is permitted. The skins yellow. This breed is little heavier than Jersey. Heifers are generally bred at the age of 17 to18 months. The milk was primarily used for butter as the milk colour is more yellower than the Jersey milk due to higher carotene content. The milk fat and SNF percentages are slightly lower than Jersey milk. The udder is less symmetrical than Jersey. Cows are active and alert but not nervous and can be easily maintained. The birth weight of calves in this breed is slightly more than that of the Jersey breed. Guernsey heifers mature slightly later than the Jersey heifers. The milk fat is nearly 5 %. Individual's cows have given 14,562 kg of milk in 365 days. Average birth weight of calves is 34 kg. Cow's weight is about 455 to 545 kg and bulls weight 727 kg. The average milk yield of cows is 2909 kg per lactation with 5% fat.

Ayershire

The home of this breed is Ayer in Scotland. These animals are distributed all over the world. The colour of the animals is red or white, with markings or, with white spottings. The red colour may be very light to almost red. The animals are beautiful with shortest top lines, levele drumps, and good udders. Horns are long and turned upwards. The animals are alert and active and they are good grazers. The average weight of the females is 455 kg and for males it is from545 to 682 kg. The calves born are strong vigorous and easy to rise and their birth weight is 32to 36 kg. The average milk yield of cows is 3664 kg with 4% fat in lactation. Individual animal has given 14,625 kg of milk in 300 days.

Crossbred Cattle Strains

Frieswal

Friesian x Sahiwal crossbreeds with Friesian inheritance between 3/8 and 5/8 are being interbred using semen of 5/8 Friesian crossbred bulls into a breed development program. The new breed has been named 'Frieswal.' The 5/8 Friesian bulls have been produced through nominated mating using proven semen of Holstein Friesian bulls on 3/8 Friesian crossbreeds. The averages for the total and 300 first lactation milk yield were 2729.9kg and 2629.5kg with a lactation length of 326.0 days. The peak yield was 12.2 kg. and average weight at first calving was 381 kg.

Jersindh

Red Sindhi x Jersey crosses had the most desirable traits for Indian conditions. These include small body size, better adaptability and high fat percentage. The Jersey crossbreeds between 3/8 and 5/8 have been interbred and named as 'Jersindh'. Similarly, 3/8-5/8 Brown-Swiss x Red Sindhi crosses have been interbred and named as 'Brown-Sind'. Jersindh crosses gave milk yield between 1557 and 1861 kg in first lactation. The breed has shown deterioration over the years mainly because of small numbers and being confined to the Institute farm.

Karan Fries

This breed has been evolved through crossbreeding between Tharparkar and Holstein-Friesian at the NDRI, Karnal, India. The breed has 50 per cent inheritance from Friesian, and is extremely docile. The average age at first calving is 30 to 32 months and the milk production is around 3700 kg with 3.8 to 4.0 per cent fat. The inter calving period is 400 to 430 days.

Karan Swiss

Karan Swiss evolved from crossing American Brown Swiss bulls with Sahiwal and Red Sindhi cows. Brown-Swiss inheritance is around 50 per cent. The colour of the breed is red dun. It resembles Sahiwal in its body size and general appearance, and the dewlap is pendulous as in the case of Sahiwal. The hump is almost non-existent; the barrel is long and deep, the naval flap is from tight to slightly loose. Eyes are full, ears small, oblong and hairy from the inside. The neck is of medium size. The legs are proportionate in size and well set apart. The udder is of good size, wide, deep and long. The udder is mostly bowl shaped; teats are cylindrical pointed or round and are of medium size. The milk veins are well developed and tortuous. The males have powerful shoulders. The average age at first calving was 32 months; the first lactation yield was 2564.7 kg with 4.2 to 4.4 per cent fat. The total milk yield based on pooled lactations was 3257.3kg with an overall calving interval of 395.5 days.

Establishment of New Breeds

The need of development of new breeds arose only when we are not satisfied with the existing breeds as regards to their utility value and when we feel that their value can be enhanced by making new gene combinations from the breeds which is likely to complement each other in different traits. In this process of formation of new breed, the various factors contributing to their development should be followed carefully. Usually we try to combine desirable traits of two or several breeds. This is done by crossing original breeds to create a heterogeneous population from which, those animals are selected with desired combination of traits. The some of the breeds developed by the breeders around the world are given below.

Synthetic Exotic Cattle Breeds Evolved

Santa Gertrudis

Santa Gertrudis were developed on King Ranch to function in hot, humid, and unfavorable environments. The Santa Gertrudis was developed by crossing Indian Brahman cattle with British Shorthorns. In 1920, years of experimentation culminated with the birth of Monkey, a deep red bull calf. Monkey became the foundation sire for not just a superior line of cattle, but an entirely new breed. In 1940, Santa Gertrudis was recognized by the U.S. Department of Agriculture as the first beef breed developed in the United States. Even today Santa Gertrudis are referred to as America's original beef breed. Santa Gertrudis cattle are approximately five-eighths Shorthorn and three-eighths Brahman. They are a deep cherry-red color with a relatively high degree of both heat and tick resistance. Santa Gertrudis females are known for their exceptional maternal traits. Their characteristics include ease of calving, good mothering ability and abundant milk supply.

Brangus

The Brangus breed was developed to utilize the superior traits of Angus and Brahman cattle. Their genetics are stabilized at 3/8 Brahman and 5/8 Angus. The combination results in a breed which unites the traits of two highly successful parent breeds. The Brahman, through rigorous natural selection, developed disease resistance, overall hardiness and outstanding maternal instincts. Angus are known for their superior carcass qualities. They are also extremely functional females which excel in both fertility and milking ability. The breed has proven resistant to heat and high humidity. Under conditions of cool and cold climate they seem to produce enough hair for adequate protection. The cows are good mothers and the calves are usually of medium size at birth. The cattle respond well to conditions of abundant feed but have exhibited hardiness under poor conditions.

Synthetic Dairy Cattle Breeds Evolved in India

Taylor Breed

This breed was evolved from a crossbred population in and around Patna. It originated in crosses between local cattle and four Shorthorn (or perhaps Channel Island) bulls introduced in 1856 by then Commissioner Mr. Taylor. No conscious selection was made among crossbreds but they were better milkers (6 to 8 kg milk / day) than the local cattle.

Karan Swiss

This breed was developed at NDRI, Karnal from crosses made between mainly Sahiwal cows and Brown Swiss bulls from Switzerland and the USA. Both F_1 and F_2 generations were obtained and a small number of cows with ¾ Brown Swiss blood were also bred. From these, females were selected and foundation stock had ½ to ¾ Brown Swiss inheritance and the breed was established in 1977.

Karan Fries

This breed was originated in a cross between Holstein Friesian bulls and Tharparkar cows at NDRI, Karnal. At the same time, some Tharparkar cows were crossed with Jersey and Brown Swiss bulls and they were backcrossed to Holstein. Subsequently ¾ exotic crosses were bred to select F_1 bulls to produce cows with ½ to 5/8 Friesian blood and were named as Karan Fries. The age at first calving and milk yield vary between 32 to 35 months and 3500 to 4000 kg respectively.

Sunandini

Sunandini was developed from the Brown Swiss crosses with local cattle generated by the Indo-Swiss Project, Mattupatti, Kerala. It was evolved by inter se mating of 5/8 Brown Swiss and 3/8 local. It was recognized as a breed in 1979. The major component of zebu in the Sunandini breed is the local nondescript cattle of Kerala. Originally conceived as a multipurpose breed for milk, draught and meat, it is now becoming solely a milk breed. The only phenotypic characteristic that can now be considered typical to the Sunandini is the straight back and comparatively short flat head. The colour varies from different shades of grey to brown. Mature body weight – 350 to 400 kg; age at first calving – 28 to 32 months; first lactation milk yield – 2300 to 2700 kg; overall lactation yield of 3200 kg; milk fat percentage of 4.

Classification of Buffalo Breeds of India

The domestic or water buffalo (*Bubalus bubalis*) belong to the family bovidae, sub-family ovinae, genus bubalis and species arni or wild Indian buffalo. Buffalo are believed to have been domesticated around 5000 years ago in the Indus Valley. The water buffalo can mainly be classified as river and swamp type. The domestication of swamp buffalo took place independently in China about 1000 years later. The movement of buffalo to other countries both east and westwards has occurred from these two countries. Some of the well-known dairy breeds of buffalo found in India and Pakistan are Murrah, Nili-Ravi, Kundi, Surti, Jaffarabadi, Bhadawari, Mehsana, Godawari and Pandharpuri. Despite potential advantages, little attention has been paid to buffalo improvement programs.

The buffalo is known as water buffalo. There are number of buffalo breeds in India but true to type and descriptive breeds are Murrah, Jaffrabadi, Nili-Ravi and Mehsana.

Buffalo breeds are classified as

Types of Breeds

1. Riverine type (or) Water buffaloes and
2. Swamp type

Murrah Group	Murrah, Nili-Ravi and Kundi
Gujarat Breeds	Surti, Mehsana and Jafarabadi
U.P. breeds	Bhadawari and Tarai
Central Indian varieties	Nagpuri, Pandharpuri, Manda, Jerangi, Kalahandi and Sambalpur
South Indian breeds	Toda and South Kanara

Important Buffalo Breeds are :

Murrah	Toda
Nili-Ravi	Pandharpuri
Jaffarabadi	Marathwada
Mehsana	Banni
Surti	Godavari
Nagpuri	Kundi
Bhadawari	

Murrah

The breeding tract of Murrah breed is Rohtak, Hisar and Jind districts of Haryana and Nabha and Patiala districts of Punjab but the animals are distributed through out India. The animals are noted for milk and fat production. The breed has a massive body; neck and head are comparatively long, horns short and tightly curved. The hips are broad, and fore – and hindquarters drooping. The tail is long reaching up to the fetlocks. The colour is usually jet black with white markings sometimes found on tail, face and extremities. The skin is soft and smooth. Ears are small, thin and pendulous. The udder is well developed with prominent veins and good sized teats. The average milk yield per lactation is 1500 to 2500 and the milk fat percentage is about 7-9 %. On an average the daily milk yield is found to be 8-10 Kg., whereas a cross breed Murrah buffalo gives 6-8 Kg. per day. The age at first calving is 45-50 months in villages but in well managed herds it is 36 to 40 months. The intercalving period is 450-500 days. The bodyweight of an adult female ranges from 430 to 500 kg and that of a male 530-575 kg.

Nili Ravi

The breed is native of Sahiwal District of Pakistan and is also found in Sutlej valley in Ferozpur District of Punjab state (India).The name Nili-Ravi comes from the supposedly blue waters of Ravi and Sutlez rivers. Animals of this breed are distributed all over India andPakistan. Skin and hair colour of Nili-Ravi are usually black although brown is not uncommon. Wall eyes and white markings on forehead, face, muzzle, legs and tail switch are the important physical features of this breed. The animal is large sized like Murrah. The frame is medium sized. The horns are small, coiled tightly and circular in cross-section. The neck is long, thin and fine. The navel is very small. The tail is long touching the ground. The udder is well developed. The milk yield is 1500 to 2300 kg per lactation. The intercalving period is 500 to 550 days. The age at first calving is 45 to 50 months. The bullocks are good for heavy trotting work. The average adult female and male weigh around 450 and 600 kg.

Bhadawari

This breed is found in Agra and Etawah districts of Uttar Pradesh and Gwalior district of Madhya Pradesh. The body is of medium size and of wedge shape. The head is comparatively small, the legs are short and stout, the hooves black, the hindquarters uniform and higher than the forequarters. The tail is long, thin and flexible with black and white or pure white markings reaching up to fetlock. The body is usually light or copper coloured which is peculiar to this breed. The udder is not well developed and teats are of medium size. The average milk production is 800 to 1000 kg with very high fat content (10-13 per cent). The bullocks are good draft animals with better heat tolerance. The average adult body weight of female and male is 386 and 476 kg, respectively.

Godavari

Godavari breed has been developed from crosses of native buffalo with Murrah bulls. The breed has attained uniformity and almost reached the production level of Murrah. The home tract of this breed is Godavari deltaic areas in Andhra Pradesh, India. The animals are medium sized with compact body. The colour is predominantly black with a sparse coat of coarse brown hair. The head is clean-cut with lean face, convex forehead and prominent bright eyes. The horns are short, flat, curved slightly downward, backward and then forward with a loose ring at the tip. The chest are deep with well-sprung ribs. The barrel is massive and long with straight back and a broad level rump. The udder is medium in size. The tail is thin and extends below the hocks with or without a white switch. Godavari buffalo are reputed for high fat with daily average milk yield of 5 to 8 kg and lactation yield of 1200 to 1500 kg. The best animals even produce around 2000kg in lactation. The animals breed regularly and have a short calving interval compared toMurrah. They are hardy and possess good resistance against the majority of the prevailing diseases.

Jaffarabadi

This breed is found in Gir forests of Kathiawad and is mainly concentrated in Kutch, and Jamnagar districts of Gujarat State. The body is long, massive and fatty. They are very massive animals with large body size requiring large quantities of fodder. The dewlap in females is somewhat loose and the udder is well developed. The head and neck are massive. The Fore head is very prominent. The horns are heavy, inclined to droop at each side of the neck and then turning up at points into an incomplete coil. The colour of the animals is black, with white patches on face and legs. The milk yield as well as milk fat is high. The average milk yield is 1000 to 1200 kg. Some good specimens yield up to 2500 kg of milk in lactation. The bullocks are heavy and used for ploughing and are also good for heavy road work. The average adult weight of female and male is 454 and 590 kg, respectively. Under optimum inputs, they may weigh 800 and 1000 kg.

Kundi

The word Kundi means fish-hook in Sindhi language. Ware (1942) first described Kundi as a distinct breed while admitting that some breeders still consider it a geographical type of Murrah.

The Kundi breed is distributed in the forest tract along the river, Indus, in the rice-growing region of north Sindh and in the swampy and rice tracts of Karachi and Hyderabad districts of Pakistan. Although, Kundi has been described as a distinct breed, some breeders still considered it a geographical variant of Murrah. Kundi animals are generally jet black (85 per cent) although light brown are not uncommon (15 per cent). Horns are thick at the base, inclined backward and upward and end in a moderately tight curl. The forehead is slightly prominent, the face hollow and eyes are small. Hind quarters are massive. Mammary glands are capacious with prominent milk veins; teats are squarely placed. Kundi buffalo are smaller than Nili-Ravi with adult weight of 320-450 kg.

Nagpuri

The Nagpuri buffalo is a versatile breed of the Vidarbha region of Maharashtra State and stands better amongst the breeds of buffaloes which combine the milk and drought qualities in a better proportion in adverse climatic conditions. The animals of this breed are very well adapted to the harsh-semi-arid conditions of Vidarbha region and can withstand extreme climatic conditions as high as $47°$ C even in respect of milk production and fertility. This breed derives its name from Nagpur district and is popularly called Varhadi (Berari), Ellicpuri / Achalpuri, particularly in Akola, Amravati, Buldhana and Yavatmal districts. The traditional breeding tract of this breed covers an area of 41,105 sq. km. and a considerable part of the tract is semi-arid and suffers acute water storage. The maximum temperature ranging between 46-47°C is not uncommon during the peak summer seasonThe natural breeding tract of the breed is Ellichpur (Achalpur), Paratwada, Daryapur and Anjangaon-Surji tehsils of Amravati districts. Typical specimens of Nagpuri buffalo are also found in the vicinity of Arvi tehsil of Wardha district. These animals in their pure form are found in Degma and Kavdas villages of Hingna tehsil; Kathlabodi and Rohna villages of Katol tehsil of Nagpur district. Jamwadi, Kalamb, Chaparda, Ghoti and Jamb-bazar villages in Yavatmal district are the main pockets of this breed. People of Nanda-gawalis and Gosavi communities are found to own this breed. Occupying isolated hamlets in remote villages, they claim to be descendants of cow-herd friends of Lord Krishna. Nanda-Gawalis rear only Nagpuri buffaloes.

Toda

Among Indian breeds of buffaloes, the Toda buffalo is a unique breed and a genetically isolated population, confined to the Nilgiri hills of Tamilnadu. These buffaloes are reared mainly by the Toda tribes who are among the most aboriginal inhabitants Toda buffaloes are ash grey coloured. Color of the calf is generally fawn at birth which varies from grey, light grey and dark grey. In growing calves, at about 2-3 months, the fawn color changes to ash grey. The horns are quite large, set wide apart, outward, and upward to form characteristics semi circle. The Crescent-shaped horns with sharp tip and two chevron markings in the neck region, are the distinguishable features of Toda buffaloes. Tail is long and slim extending beyond hock joint and the switch is generally black. Toda buffaloes are medium in size and are considered to be quite powerful. Body is fairly long with a broad and deep chest. Head is large heavy, carried to the level of body. Udder is small and not so prominent of this country. Various values of morphological traits of Toda buffaloes like body length, height at withers, heart girth and adult weight are given as, Body Length (cm) 133 ± 0.10 ,Body Length (cm) 133 ± 0.10, Body Length (cm) 133 ± 0.10, Height at withers (cm) 122 ± 0.60, Hearth Girth (cm) 180 ± 1.10.

Swamp

The swamp buffalo are concentrated mainly in south east China, Myanmar, Malaysia, Laos, Cambodia, Thailand, Indonesia, Philippine, and Vietnam. The skin colour is gray, dark gray to state blue. White animals occur frequently. Animals have swept back horns and are similar in appearance across the countries except the size. The horns grow laterally and horizontally in young animals and curve round in a semi circle as the animals gets older. Animals are massively built, heavy bodied with large belly. The forehead is flat; orbits are prominent with a short face and wide muzzle. They weight from 300 to 400 kg when fully grown. Swamp buffalo are primarily used as work animal

Some villages also provide artificial insemination. The average milk yield per animal per day in Mehsana buffaloes ranges from 4.37 to 4.81kg. However, a systematic Mehsana breed improvement programme through field progeny testing was launched in 1985 in the milk shed area of the Mehsana district. 107 bulls were tested in eight batches. Average 305 day first lactation milk yield of 50 daughters of the top proven bulls of the first four batches in these buffaloes ranged from 2 085 to 2 312 kg.

Nagpuri

It is an improved local breed, the result of a selection of Indian breeds of buffaloes.

Black in colour, sometimes there are white markings on the face, legs and switch. Horns are 50-65 cm long, flat-curved and carried back near to the shoulders. Nasal flap is mostly absent and even if present is very short. Height at withers of adult male is 140 cm, body weight is 522 kg. Height at withers of adult female is 130 cm, body weight is 408 kg., Lactation duration 243 days, Milk yield 825 kg, Milk fat 7.0 percent. This breed is raised in the Nagpur, Wardha and Berar districts of Madhya Pradesh.

Buffaloes are traditionally managed under domestic conditions together with the calf. They are hand-milked twice a day. They are fed different kinds of roughages: barley and wheat straw, cornstalks, sugar cane residuals. In addition, they are given concentrate mixtures. If grazing is available, they graze all day long. They are naturally mated.

Sambalpuri

This breed is raised around Bilaspur in Chhattishgarh (India). Black in colour, with white switch on tail, with narrow and short horns, curved in a semi-circle, running backward, then forward at the tip.

Buffaloes are traditionally managed under domestic conditions together with their calf. They are hand-milked twice a day. They are fed different kinds of roughages: barley and wheat straw, cornstalks, sugar cane residuals. In addition, they are given concentrate mixtures. If grazing is available, they graze all day long. They are naturally mated. It is a good healthy draught animal with a rapid pace and it is comparatively the most productive breed of the region. Some exceptional buffaloes may yield as high as 2 300 to 2 700 kg in about 340 days. Lactation duration 350 days. Milk yield 2400 kg.

in paddy cultivation, for pulling carts and hauling timber in jungles. Milk yield is 1-2 kg per day.

Manda

These buffaloes are found in whole Koraput district and adjoining parts of Malkangiri and Nawarangpur district in Orissa, spread over an area of around 10,000 sq. km. Body colour of these buffaloes is ash grey and grey with copper coloured hairs. Some animals are silver white in colour.

The lower part of the legs up to elbow is light colour with tuft of yellowish/copper coloured hairs at knee and fetlock. Horns are broad and emerge slightly laterally, extending backward and inward making half circles. Jaws and nostrils are wide and prominent.

Manda buffalo comes to heat at around 40 months and drops its first calf at around 51 months of age. Average calving interval of these buffaloes is 18 months with gestation period of 307 days. These buffaloes are moderate milk yielders having lactation milk yield of around 700 lt. in a lactation length of 290 days. Age at first calving 1534.43±3.76days. These animals are famous for longevity, hard work and length of working life. Female Manda buffaloes at some places are used in agricultural operations along with buffalo bullocks.

Mehsana

The existence of the Mehsana breed in north Gujarat, India, is referred to in 1940. This breed is the result of selection of Indian breeds of buffalo. This breed Concentrated between the Mahi and Sabarmati rivers in Gujarat (India).

Characteristics are intermediate between Surti and Murrah. Jet black skin and hair are preferred. Horns are sickle-shaped but with more curve than the Surti. The udder is well developed and well set. Milk veins are prominent. Body weight of adult male is 570 kg.

Body weight of adult female is 430 kg. Lactation duration 305 days ,Milk yield in a lactation 1800-2 700 kg,Milk fat 6.6-8.1 percent, Milk protein 4.2-4.6 percent, Products.

Buffaloes are traditionally managed under domestic conditions together with the calf. They are hand-milked twice a day. They are fed different kinds of roughages: barley and wheat straw, cornstalks, sugar cane residuals. In addition, they are given concentrate mixtures. If grazing is available, they graze all day long. They are naturally mated.

Surti

The existence of the Surti breed in north Gujarat (India) is referred to in 1940. It is the result of a selection of Indian breeds of buffalo. It is one of the most important breeds in Gujarat and in Rajasthan.

Coat colour is black, skin is black or reddish. They have two white chevrons on the chest. Animals with white markings on forehead, legs and tail tips are preferred. Horns are flat, of medium length, sickle shaped and are directed downward and backward, and then turn upward at the tip to form a hook. The udder is well developed, finely shaped and squarely placed between the hind legs. The tail is fairly long, thin and flexible ending in a white tuft. Height at withers of adult male is 131 cm; body weight is 700 kg. Height at withers of adult female is 124 cm; Average body weight (kg) 462±7.0, Age at first calving (months) 53.2±1.7, First lactation 305 days or less yield (kg) 1 295±57, All lactation 305 days or less yield (kg) 1 477±42, All lactation total yield (kg) 1 547±50, All lactation length (days) 311±7, Average fat (percent) 8.10, Average dry period (days) 234±21, Service period (days) 207±17, Calving interval (days) 510±16, Number of services per conception 2.55, Milk fat 6.6-8.1 percent, Milk protein 4.2-4.6 percent.

Concentrated between the Mahi and Sabarmati rivers in Gujarat (India).

Buffaloes are traditionally managed under domestic conditions together with the calf. They are hand-milked twice a day. They are fed different kinds of roughages (barley and wheat straw, cornstalks, sugar cane residuals). In addition, they are given concentrate mixtures. If grazing is available, they graze all day long. They are naturally mated. Some villages also provide artificial insemination.

Pandharpuri

The Pandharpuri buffaloes are known to have been reared for more than 150 years in the breeding tract. The local "Gawli" community reared these buffaloes for milk production. These buffaloes had royal patronage from Kolhapur for supply of fresh milk to the wrestlers of Kolhapur. The breed is famous for its better reproductive ability, producing a calf every 12-13 months. Under average management conditions and hot-dry climate, these buffaloes yield 6-7 litres of milk per day. However, under good management they are reported to yield up to 15 litres of milk in a day.

The main breeding tract of Pandharpuri buffalo is Solapur, Sangli and Kolhapur districts of Maharastra state which is mostly drought prone. The soil of is black, coarse gray and reddish. In Sangli it is black lateritic, saline, alkaline in low lying patches, while it is black, red-lateritic and black-brown in Kolhapur districts.

The breeding tract of Pandharpuri buffalo comes under scarcity, plain and sub-mountain agro-climatic zones. Average minimum and maximum temperature is between 9°C to 42°C in the breeding tract. The average minimum and maximum humidity is between 43 to 87 percent. The annual rainfall varies between 345.64mm in Sangli to 1168.96mm in Kolhapur district. The rainfall occurs between middle of May to end of October with most rainfall occurring during June to September. The fodder crops commonly grown include Jawar, Maize, Bajra, Oat, Lucerne, Berseem etc. Body colour Black in 86.51% and brown in 13.14% animals. The horns are very long, running backwards, upwards and twisted outward and touching almost backbone. Four types of horn orientation described which are locally called as Toki (52.05%), Bharkand (34.24%), Meti (10.81%) and Ekshing meti (2.09%). In majority of Pandharpuri buffaloes (49.92%), horn tip directed upward while in 28.23% buffaloes they are lateral White spots may be found on forehead, muzzle and pastern regions of body.

Tail is short, white switch is common and the colour pattern of the switch of tail comprises 72.92% white and 27.08% black. Medium sized and compact body animal and hair colour is grey, tan and blakish. Forehead is convex in (92.39%) and flat in(7.61%) buffaloes. Long and narrow face, prominent nasal bone and comparatively narrow frontal bone, comparatively longer and thin. Ear orientation is horizontal (60.51%) and dropping (39.49%) in animals. "Trough udder" is found in 52 to 56% buffaloes followed by "Bowl udder" in 34 to 36%. The "Pendulous udder"(5.6 to 7.2%) and "Round udder"(8 to 10 %) is negligible in this breed. Most of the Pandharpuri buffaloes have cylindrical type of teat (47.87%).

Age at First Calving (months) 43.82, Service Period (days) 97.92, Calving Interval (days) 07.05, Average lactation length (days) 255.60 + 14.7, Average lactation milk yield (kg) 1207.70 + 13.4, Average daily milk yield (kg) 4.90+0.08, Average Fat % 7.80 + 0.07.

Banni

Banni Buffalo breed was recognized as 11th buffalo breed of India by Breed Registration Committee, ICAR, New Delhi. The breed is originated from the Banni area of kachchh, which is a part of Kachchh district of Gujarat. Purebred animals prevalent in Bhuj, Nakhatrana, Anjar, Bhaahau, Lakhpat, Rapar and Khavda talukas, are heavily size with typical double and vertical coiling of the horn.The total area of Banni grassland is about 3847 sq. km.

The *"sui-genesis"* germplasm of Kachchh i.e. *"Banni buffaloes"* are maintained by maldharis under typically and locally adapted extensive production system in its breeding tract. The breed is very hardy, well adapted to harsh climatic conditions. The performance recorded under field conditions shows the potentiality of the animals, which is regular in breeding and have high milk production potential. Banni buffalo known for high productivity, hardiness, reared under extensive production system, thrives on grasses available in the Banni grass land through night grazing and only source of livelihood for landless maldharis. In Banni, Maldharis adopted animal husbandry exclusive livelihood approach; they have been invented locally adapted extensive production system to reduce the cost of production. Banni buffaloes are trained to typical grazing on banni grass land during night and come to the villages in the morning for giving milk. This traditional system of buffalo rearing has been adapted to avoid the heat stress and high temperature of the day. The body coat colour is black (90.09%) and copper (9.90%), whereas muzzle and eyelids are either black or brown Horns orientation is vertical, inverted double coiling in 31.20% and vertical, inverted single coiling in 68.80% animals Eyes are prominent black and bright. Age at First Calving 40.28+0.25 (months) Service Period (days) 81.77+ 4.41 Calving Interval (days) 85.64+ 4.23 Dry Period (days) 372.43+ 3.97 Average lactation length (days) 300.96 + 4.43 Average lactation milk yield (kg) 2857.21 + 89.76 Peak milk yield (kg) 14.87 + 0.21 Average Fat % 6.65 + 0.11

The animals of this breed were also taken to the adjoining state i.e. Madhya Pradesh, Maharashtra and Rajasthan by livestock breeders. The colour pattern of the switch of tail comprises (67.35%) white and (32.65%) black and length of the tail is 88.39+-0.48cm. Medium to large, compact and generally covered with hairs. Dewlap is absent and naval flap is medium. Well developed, round in shape and squarely placed. The hind and fore quarters are uniformly well developed, whereas typically whole udder looks like four equal divisions with teats well attached to each quarter.

Classification of Sheep Breeds

Sheep (domestic group)	
Phylum	Cordata
Sub-phylum	Craniata (Vertebrata)
Class	Mammalia
Sub-class	Theria (Viviparous)
Infra-class	Eutheria (Placenta)
Order	Ungulata (hoofed mammals)
Sub-order	Artiodactyles (even-toed)
Sub-division	Pecora (true ruminants)
Family	Bovidae (hollow -horned)
Genus	Ovis
Species	Aries

Sheep Breeds of India

The country has 40 breeds of sheep out of which 24 are distinct. Amongst them 5 can be classified into medium or fine wool, 14 into coarse carpet quality wool and rest into hairy meat type breeds.

They vary from the non-woolly breeds of sheep in the Southern peninsular region mainly kept for mutton and manure to reasonably fair apparel wool breeds of Northern temperate region.

Although productivity from these sheep is of low order, they cannot altogether be considered inefficient in comparison to the physical, environmental and nutritional conditions they are reared in.

If we follow the breed classification in strict sense, there are no specific breeds, as majority of them lack characteristic of fixed nature.

Neither are there breeding societies, nor agencies to register animals of particular breeds, maintain flock books and ensure purity of the breeds. The sheep from various states keep on migrating over long distances where they undergo lot of admixture and thus making it difficult to maintain purity of the breeds.

Animals with distinct characters localised to a place and different from those of other place are termed as breeds and given some local name.

There have been little efforts to conserve and improve the native breeds except for at some of the Central and State Government farms.

Some important breeds of sheep are maintained for pure-breeding and producing stud rams for distribution to the farmers. Most of the breeds of sheep in India have evolved through natural adaptation to agro-ecological conditions, followed by some limited artificial selection for particular requirements.

Most of the breeds have generally been named after their place of origin and on the basis of prominent characters.

Amongst the most widely distributed native sheep breeds, Marwari and Deccani are the most prevalent; out of them Marwari covers the greater part of arid North-western region of Rajasthan and Gujarat. It is highly migratory, following a trans-human system of management and has left greatest impact on other breeds, especially those with very coarse and hairy fleece like Malpura and Sonadi. Deccani occupies most of the Central part of the Southern peninsula being distributed over the states of Maharashtra, Andhra Pradesh and Karnataka.

Attempts have been made to define and document some of the important breeds of sheep under ad-hoc research scheme financed by the Indian Council of Agricultural Research. Such efforts were mostly based on the exterior phenotype, shape and length of ears, length and direction of horns, fleece type, body colour and tail length, etc.

There was little serious consideration to body weight, body measurements, population and flock size and its structure, prevalent management practices, productivity status and problems associated with their conservation and further development.

The efforts of Acharya, (1982) to define some of the existing breeds are based on consideration of agro climatic regions and the type of sheep found therein and adult body weights and linear biometry on representative samples of sexes, and production performance, both published and unpublished and personal surveys.

Based on variable agro-climatic conditions and over places and type of sheep found in them, the following four different regions regrouped from 15 agro-climatic zones are distinguishable over the country

North Temperate	North-Western Arid and Semi Arid	Southern Peninsular	Eastern
Bhakarwal (CW)	Chokla (CW)	Bellary (MCW)	Balangir (MCW)
Changthangi (CW)	Hissardale (AW)	Coimbatore (MCW)	Bonpala (MCW)
Gaddi (CW)	Jaisalmeri (MCW)	Daccani (M)	Chottanagpuri (MCW)
Gurez (CW)	Jalauni (MCW)	Hassan (M)	Ganjam (MCW)
Karnah (AW)	Kheri (MCW)	Kanguri (M)	Garole (M)
Kashmir (AW) Merino	Magra (CW)	Kilakarsal (M)	Tibetan (CW)
Poonchi (CW)	Malpura (MCW)	Madras Red (M)	
Rampur (CW) Bushair	Marwari (MCW)	Mandya (M)	
	Muzaffar-nagari (MCW)	Mecheri (M)	
	Nali (CW)	Nellore (M)	
	Patanwadi (CW)	Nilgiri (AW)	
	Pugal (MCW)	Rammand White (M)	
	Sonadi (MCW)	Tiruchy Black (M)	
	Munjal (M)	Vembur (M)	

Within parenthesis is the major product of the breed. (AW), Apparel wool; (CW), Carpet wool; (MCW), Mutton and Carpet wool; (M), Mutton.

Details of some of the important breeds are given below:

Avikalin

Avikalin strain has been evolved from Rambouillet x Malpura half-bred base through sinter breeding and selection for greasy fleece weight, producing about 1.75 kg greasy wool having 27 microns diameter, 27% medullation and staple length of 4.75 cm in half yearly clip. This breed is quite suitable as a dual purpose sheep for carpet wool and mutton production.

Balangiri

This breed is spread over north western districts of Orissa, Balangir, Sambalpur and Sundargarh. They are medium sized, white or light brown or of mixed colours. A few animals are black. The ears are small and stumpy. Males are homed and females are polled. The tail is of medium length and thin. The legs and belly are devoid of wool. Fleece is extremely coarse, hairy and open.

Bellary

Distributed in Bellary district of Karnataka. Sheep found to the north of the Tungabhadra River are called "Deccani" and those to the south of it are called as, "Bellary". Medium-sized animals, with body colour ranging from white through various combinations of white and black to black. Fleece is extremely coarse, hairy and open; belly and legs are devoid of wool. Means of body weights at birth, three, six and twelve months of age were 2.60, 11.09, 16.28 and 18.68 Kg, respectively. Average 6-monthly greasy fleece weight was 300g with average fiber diameter of 59.03 micron and Medullation percent of 43.43. Litter size is single with rare case of twinning.

Bharat Merino Sheep

Bharat Merino sheep developed by crossbreeding indigenous Chokla and Nali sheep with Rambouillet and Merino rams and stabilised at 75% exotic inheritance has the potential as an import substitute for exotic fine wool inheritance. The annual greasy wool production is 2.5 kg with fibre diameter of 19 20 microns, medullation less than one percent and staple length in annual clip at Mannavanur of 7 8 cm. From SRRC, Mannavanur Bharat Merino sheep were distributed to Kolar District of Karnataka and in Sathyamangalam and Thalavadi areas of Erode District in Tamilnadu. The average body weight of Bharat Merino sheep at Kolar District was found to be 45 kg and 32 kg, ranged from 38-70 kg and 28-40 kg for adult male and female respectively. Birth weight of lamb was 4.0 kg with range of 3.5-4.5 kg. The weight at 3, 6, and 12 months were 18, 25 and 30 kg, respectively.

Bonpala

This breed is found in southern Sikkim. The animals are tall, leggy and well built. Fleece colour ranges from completely white to completely black with a number of intermediary tones. Ears are small and tubular. Both sexes are homed. The tail is thin and short. The belly and legs are devoid of wool. Fleece is coarse hairy and open. Average six monthly greasy fleece yield is 500 g. Fiber diameter is 49 µ with Medullation percent of 70.5.

Chokla

This breed is distributed in Churu, Jhunjhunu, Sikar and bordering areas of Bikaner, Jaipur and Nagaur districts of Rajasthan. Animals true to the breed type are found in Sikar and Churu districts. The animals are light to medium sized. The face, generally devoid of wool, is reddish brown or dark brown, and the colour may extend up to the middle of the neck. The skin is pink. The ears are small to medium in length and tubular. Tail is thin and of medium length. The coat is dense and relatively fine, covering the entire body including the belly and the greater part of the legs. Chokla is fine carpet wool Indian sheep and reared basically for its wool quality and suitability for migration. Chokla grows the finest carpet wool of all the Indian breeds ranging in its quality number from 54s to 60s count. The wool produced by Chokla sheep is heterogeneous and is generally mixed with coarser fleece of other sheep before utilization as carpet wool. Over the period there is significant improvement in the body weight at different ages and this reflects the effect of selection in the flock. Being a Best Carpet wool breed, Chokla produces wool with average fiber diameter and medullation percentage of around 30m and 30% with staple length of more than 6.0 cm suitable for all kind of carpet preparation. Since 1992, through intensive selection and improved management, six month weight of the sheep has increased significantly from 16.51 kg to 24.83 kg and first 6 monthly GFY increased from 0.918 kg to 1.438 kg. A total of 300 rams were sold / distributed to the farmers or to the Govt. of Rajasthan for breed improvement programme.

Chottanagpuri

This breed is distributed in Chottanagpur, Ranchi, Palamau, Hazaribagh, Singhbhum, Dhanbad and Santhal Parganas of Jharkhand, and Bankura district of West Bengal. Animals are small, light in weight with light grey and brown colour. Ears are small and parallel to the head. Tail is thin and short. Fleece is coarse, hairy and open and is generally not clipped.

Coimbatore: Distributed in Coimbatore district and adjoining Dindigul district of Tamil Nadu. Medium-sized animals, white with black or brown spots. Ears are medium-sized and directed outward and backward. Tail is small and thin Fleece is white, coarse, hairy and open. The animals are maintained on grazing without any supplementation. Sheep are penned in harvested fields. Rams are both horned and polled while ewes are hornless. Average body weights at birth, three, six and twelve months of age were 2.16, 7.50, 10.83 and 14.77 Kg, respectively. Average 6-monthly greasy fleece weight was 365g with average fibre diameter of 41 micron and Medullation percent of 58.

Deccani

The Deccani breed of sheep is widely distributed in the Deccan plateau across the three states of Maharashtra, Andhra Pradesh and Karnataka. The breed has a thin neck, narrow chest, prominent spinal processes. It has Roman nose and dropping ears. The colour is dominantly black, with some grey and roan. Different strains (or within breed types) are observed in the breed tract. Four types have been noticed. Local people term them as Viz. Lonand, Sangamneri, Solapuri (Sangola) and Kolhapuri. Means of body weights at birth, three, six, nine and twelve months of age were 3.13, 14.30, 18.20, 20.10 and 22.57 Kg, respectively for Deccani sheep maintained under NWPSI. Sheep were sheared two times a year and the overall means for lambs first six monthly clip and second six monthly clip and adult annual clips were 467, 499 and 488 g, respectively. Deccani lambs were put on feedlot trails (90 days) after weaning with an initial average weaning weight of 13.23 kg. Final weight at six months of age was 22.65. Average gain of 9.13 kg was obtained with a with a feed efficiency ratio of 1: 5.91. Results shows that the Deccani has the great potential for mutton production under intensive system of management.

Gaddi

Gaddi sheep are distributed in Kistwar and Bhadarwah tehsils in Jammu province of Jammu & Kashmir, Ramnagar, Udampur and Kullu and Kangra valleys of Himachal Pradesh and Dehradun. Nainital, Tehri Garhwal and Chamoli districts of Uttarakhand. The animals are medium sized, usually white. Although tan, brown and black and their mixtures are also seen. Males are homed; 10 to 15 per cent females are also homed. Tail is small and thin. The overall least square means for 1st six monthly and adult six monthly were 417 and 454g respectively and least square means of birth, 3, 6, 9 and 12 month's weight of lambs were 2.22, 7.64, 11.01, 14.01 and 16.19 kg, respectively under farm conditions. Average fibre diameter and medullation were 27.7µ and 20.6 % respectively.

Ganjam

Ganjam sheep are found in Koraput, Phulbani and parts of Puri districts of Orissa. The animals are medium sized with coat colour ranging from brown to dark tan. Some animals have white spots on the face and body. Ears are of medium size and drooping. The nose line is slightly convex. The tail is of medium length and thin. While the males are horned, the females are polled. The fleece is hairy and short and not shorn. An annual lambing percentage of 83.6 and mortality percentage of 10.35 have been recorded in farmers' flocks. Under the network project on sheep improvement, field based Ganjam unit was started in year 2001 at Orissa University of Agriculture and Technology, Bhubneswar for mutton.

Garole

Sheep are reputed for multiple births and are found in the Sunderban area of West Bengal. These sheep are reported to have contributed prolificacy to the Booroola Merino sheep. Sundarban area has a sheep population of about 0.16 million. The breed tract is tropical humid. The climate of the area is hot and humid. Garole Wool is for rough carpet use. The average fibre diameter, medullation, staple length and crimp/cm. Were 67.82 µ, 75.17%, 5.09 cm and 2.08 respectively. Litter size at first lambing is two and at subsequent lambing is 2 to 3. Prolificacy reported is singles 25 30%, twins' 55 60%, triplets' 15 20% and quadruplet's 1 2%.

Hassan

Distributed in Hassan district of Karnataka. Small sized animals. Animals are having white body with light brown or black spots. Ears are medium-long and drooping; ear length. Majority of males are horned and females are usually polled. Fleece is white, extremely coarse and open; legs and belly are generally devoid of wool. This breed is very hardy and capable to travel long distances.

Hissardale

Hissardale was evolved at the Government Livestock Farm, Hissar, through crossbreeding Australian Merino rams with Bikaneri (Magra) ewes and stabilizing the exotic inheritance at about 75 per cent. There is a small flock of this breed at the Government Livestock Farm, Hissar. The rams of this breed were earlier distributed in the hilly regions of Kullu, Kangra etc.

Jaisalmeri

Jaisalmeri breed is found in Jaisalmer, Banner and Jodhpur districts of Rajasthan. Pure specimens are found in south-western Jaisalmer, extending up to north-western Barmer and southern and western Jodhpur. The animals are tall and well built with black or dark brown face and the colour extending up to the neck and typical Roman nose. Long drooping ears, generally with a cartilaginous appendage are seen. Both sexes are polled. The tail is medium to long. The fleece colour is white and fleece is of medium carpet quality and not very dense.

Kashmir Merino

This breed originated from crosses of different Merino types with predominantly migratory native sheep breeds such as Gaddi, Bhakarwal and Poonchi. The level of inheritance varies from very low to almost 100 per cent Merino, though a level from 50 to 75 per cent predominates. The animals are highly variable because of the involvement of a number of native breeds. The annual greasy fleece weight (kg) is 2.80 having average fiber diameter of 20.4μ.

Kendrapada

Sheep is identified as another prolific sheep of India after Garole of West Bengal. It is 2nd sheep breed of India which carry FecB mutation. Kendrapada is distributed in Bhadrak, Konark and Puri district of coastal area of Orissa. A survey was made to study the prolificacy of Kendrapada sheep of Orissa around Kendrapada district and Nimapara (Konark). The survey reveals that Kendrapada is a prolific and excellent medium stature meat type sheep. In the flocks surveyed more than 75 % ewes produces multiple births and the adult body weight of sheep was about 23 kg. The coat colour varies from white to dark brown. Some black animals and black/ white spot on body are also found during survey.

Kenguri

Distribution is hilly tracts of Raichur district (particularly Lingasagar, Sethanaur and Gangarati taluks) of Karnataka. Medium-sized animals. The body coat is dark brown or of coconut colour. The breed is also known as Tenguri. Some Kenguri sheep have black belly and are known as Jodka. A few sheep, known as Masaka, are a mixture of brown and black colour. Ears are medium long and drooping. Males are horned; females are generally polled. Kenguri sheep are maintained for mutton. Lactation length varies from 3 to 5 months, and the peak milk production is 300 to 500 ml/day.

Kheri

Sheep evolved in the farmers flock under the field conditions especially in the areas that are important for sheep migration in Rajasthan appears to be a need of migratory sheep breeders. The animals of this breed have established owing to their ability to walk long distances and ability to survive on limited amount of coarse feed. The animals of this breed can sustain stress and on return of favorable condition regain faster resulting in better reproduction rate and growth of lambs as compared to prevalent breeds in the area.

Kilakarsal

Distributed in Ramnathpuram, Madurai, Thanjavur and Ramnad districts of TamilNadu. Medium-sized animals. Coat is dark tan, with black spots on head (particularly the eyelids and lower jaw), belly and legs. Ears are medium sized. Tail is small and thin. Males have thick twisted horns. Most animals have wattles. Average body weights at birth, three, six and twelve months of age were 2.29, 8.53, 14.15 and 27.26 Kg, respectively.

Madras Red

Distributed in Chennai, Kancheepuram, Tiruvellore, Villupuram and adjoining areas of Vellore, Cuddalore and Thiruvannamalai districts of Tamil Nadu. Medium-sized animals. Body colour is predominantly brown, the intensity varying from light tan to dark brown; some animals have white markings on the forehead, inside the thighs and on the lower abdomen. Ears are medium long and drooping; ear length. Tail is short and thin. Rams have strong corrugated and twisted horns; the ewes are polled. The body is covered with short hairs which are not shorn. Average body weights at birth, three, six and twelve months of age were 2.71, 9.97, 14.84 and 21.03 kg, respectively. Dressing percentage was around 49.

Magra

Magra sheep is distributed in Bikaner, Nagaur, Jaisalmer and Churu districts of Rajasthan. However, animals true to the breed type are found only in the eastern and southern parts of Bikaner district.

The animals are medium to large in size. White face with light brown patches around the eyes is characteristic of this breed. Skin colour is pink Ears are small to medium and tubular. Both sexes are polled. Tail is medium in length and thin. Fleece is of medium carpet quality, extremely white and lustrous and not very dense. The average first six monthly greasy fleece yield (GFY) and adult six monthly GFY of Magra sheep under farmers conditions were 0.979 and 0635 kg Average body weights at birth, six and twelve months of age were 3.04, 20.64 and 29.23 Kg, respectively under field conditions. Fibre diameter was ranged between 32.5 to 38μ with staple length ranged from 4.2 to 6.8cm. Under farm conditions, average body weights at birth, three, six, nine and twelve months of age were 2.98, 14.77, 22.06, 25.92 and 27.93 Kg, respectively. Means for first six monthly GFY and adult annual GFY of Magra sheep under farm conditions were 0.912 and 2.071 kg respectively. Average fibre diameter, medullation and staple length were 32.55μ, 49.91 % and 6.47cm.

Malpura

Malpura sheep are found in Jaipur, Tonk, Sawaimadhopur and adjacent areas of Ajmer, Bhilwara and Bundi districts in Rajasthan. The animals are fairly well built, with long legs. Face is light brown. Ears are short and tubular, with a small cartilaginous appendage on the upper side. Both sexes are polled. Tail is medium to long and thin. The fleece is white, extremely coarse and hairy. Belly and legs are devoid of wool. The overall least square means for 1st six monthly and adult annual GFY were 551 and 810g respectively and least square means of birth, 3, 6 and 12 month's weight of lambs were 3.02, 15.41, 20.80 and 25.60 kg, respectively under farm conditions. The average fibre diameter was 41.67μ with medullation of 75.9 %. Staple length was 4.9cm.

Mandya

It is also known as Bannur and Bandur. Distributed in Mandya district and bordering Mysore district of Karnataka. Relatively small animals. Colour is white, but in some cases face is light brown, and this colour may extend to the neck. Compact body with a typical reversed U-shape conformation from the rear. Ears are long, leafy and drooping. Tail is short and thin. A large percentage of animals carry wattles. Slightly Roman nose. Both sexes are polled. Coat is extremely coarse and hairy. Evenly placed short and stumpy legs and wide apart hipbones indicated a square type meaty conformation of the breed. Sheep are mainly kept on grazing, however, a few farmers keep the animals on stall-feeding. Some of the farmers practice sheep tethering in their cropland, shifting from place to place and bringing back to their homestead in the evening. Average body weights at birth, three, six and twelve months of age as 2.02, 9.44, 14.61 and 21.34 kg, respectively. Average 6-monthly greasy fleece weight was 298g.

Marwari

Marwari sheep are distributed in Jodhpur, Jalore, Nagaur, Pali and Barmer districts extending up to Ajmer and Udaipur districts of Rajasthan and the Jeoria region of Gujarat. The animals are medium size with black face, the colour extending to the lower part of neck. Ears are extremely small and tubular. Both sexes are polled. Tail is short to medium and thin. The fleece is white and not very dense. The overall least square means for 1st six monthly, adult six monthly and adult annual were 607.16, 631.25 and 1260.50g, respectively and least square means of birth, 3, 6, 9 and 12 month's weight of lambs were 3.05, 14.74, 19.33, 22.85 and 25.90 kg, respectively under farm conditions. Average fibre diameter, medullation and staple length were 31.9μ, 50.8 % and 5.35cm respectively.

Mecheri

Distributed in Mecheri, Kolathoor, Nangavalli, Omalur and Tarmangalam Panchayat Union areas of Salem district and Bhavani taluk of Coimbatore district of Tamil Nadu. Medium-sized animals, and light brown in colour. Body is covered with very short hairs which are not shorn. The skin is of the highest quality of sheep breeds in India and is highly prized. Average body weights at birth, three, six, nine and twelve months of age were 2.23, 10.07, 13.73, 18.62 and 21.12 for male lambs and corresponding value for female lambs were 2.28, 9.70, 13.53, 15.84 and 18.02 kg, respectively under farm conditions. Corresponding growth performances recorded in farmers' flocks were 2.21, 9.64, 13.25, 16.54 and 18.72kg, respectively.

Muzzafarnagri

This breed is distributed in Muzzafanagar, Bulandshaher, Saharanpur, Meerut, Bijnor and Dehradun districts of Uttar Pradesh and parts of Delhi and Haryana. Pure specimens are found in

Muzzafarnagar district. The animals are medium to large in size. Face line is slightly convex. Ears and face are occasionally black. Both sexes are polled. Males occasionally show rudimentary horns. Ears are long and drooping. Tail is extremely long and reaches fetlock. The fleece is white, coarse and open. Belly and legs are devoid of wool. The overall least square means for lambs 1st and 2nd six monthly and adult annual clips were 582.17, 538.75 and 1217.62g, respectively under farm conditions. The overall averages were 3.63, 15.59, 23.52, 27.14 and 30.76 kg, respectively at birth, 3, 6, 9 and 12 month age. Least square means of wool quality attributes viz. fibre diameter, hetro fibres, hairy fibres, medullation and staple length were 38.39±1.36µ, 14.35±1.37%, 46.89±3.42%, 61.03±4.33% and 5.09±0.30cm, respectively.

Nali

The Nali sheep is found in Ganganagar, Churu and Jhunjhunu districts of Rajasthan, southern part of Hissar and Rohtak districts of Haryana. The animals are medium sized. Face colour is light brown and skin colour is pink. Both sexes are polled. Ears are large and leafy. Tail is short to medium and thin. Fleece is white, coarse, dense and long stapled. Forehead, belly and legs are covered with wool. The overall means for 1st six monthly and adult annual were 1.01 and 2.84 Kg, respectively and means of birth, 3, 6, 9 and 12 month's weight of lambs were 2.43, 10.74, 14.93, 17.13 and 19.64 kg, respectively under farm conditions. Average fiber diameter, medullation and staple length were 29.89µ, 41.14 % and 6.79cm respectively.

Nellore

Distributed in Nellore district and neighbouring areas of Prakasham and Ongole districts of Andhra Pradesh. Three varieties are distinguished, primarily on the basis of colour: "Palla", completely white or white with light brown spots on head, neck, back and legs; "Jodipi", white with black spots, particularly around the lips, eyes and lower jaw, but also on belly and legs and "Dora", completely brown. Relatively tall animals with little hair except at brisket, withers and breech. The rams are horned and ewes are almost always polled. The ears are long and drooping and most of the animals carry wattles. The overall averages of body weight of Nellore sheep maintained under NWPSI were 2.95, 13.91, 17.38, 22.39 and 26.61 kg respectively at birth, 3, 6, 9 and 12 month age.

Nilgiri

Nilgiri breed was evolved during the 19th century. It originated from a crossbred base and contains an unknown level of inheritance of Coimbatore, the local hairy breed, Tasmanian Merino, Cheviot and South Down. The animals are medium sized and white, exceptionally with brown patches on face and body. Face line is convex, giving a typical Roman nose. The ears are broad, flat and drooping. Males have horn buds and scurs; females are polled. Attempts have been made by Sheep Breeding Farm, Sandynallah to improve these sheep through crossing with Rambouillet and Russian merino which has resulted in substantial improvement in fleece weight and quality. Means of body weight at birth, 3, 6, 9 and 12 months of age were 2.74, 9.64, 14.06, 17.04 and 19.44 kg respectively. Average six monthly wool yields were 438g with fibre diameter of 24.3μ and staple length of 4.5cm.

Patanwadi

Patanwadi is also called desi, Kutchi, Vadhiyari and Charotari. The animals, in general, are medium to large with relatively long legs. They have typical Roman nose with brown face which may be tan in a few cases. Ears are medium to large, tubular with a hairy tuft. The tail is thin and short. Both sexes are polled. The overall least square means for 1st six monthly and adult annual GFY were 566 and 1063g respectively and least square means of birth, 3, 6, 9 and 12 month's weight of lambs were 3.23, 14.26, 17.67, 20.76 and 24.22 kg, respectively under farm conditions. Average fiber diameter and medullation were 29.11μ and 38.41 % respectively.

Pugal

Pugal area of Bikaner district is its home tract. It is distributed over Bikaner and Jaisalmer districts of Rajasthan. The animals are fairly well built. Face is black with small light brown stripes on either side above the eyes; the lower jaw is typically light brown. The black colour may extend to neck. Ears are short and tubular. Both sexes are polled. Tail is short to medium and thin. The fleece is white.

Ramnad White

Distributed in Ramnad district and adjoining areas of Tirunalveli district of Tamil Nadu. Medium-sized animals, predominantly white; some animals have fawn or black markings over the body. The ears are medium-sized and directed outward and downward. Males have twisted horns and females are polled. Tail is short and thin. Average body weights at birth, three, six and twelve months of age were 1.68, 7.31, 8.45 and 16.30 Kg, respectively.

Sandyno

Sandyno breed was developed by crossing Nilagiri sheep with Merino and Rambouillet. The breed has ¾ inheritance from Nilagiri and 5/8 inheritance from Merino / Rambouillet breeds. The breed has been field tested and found to adapt well to the local climatic and environmental conditions. The Sandyno sheep grow to a body weight of 24 kg in males and 20 kg in females at yearling stage. The average wool yield is 2.2 kg. A flock of 320 sheep of Sandyno breed is maintained at Sheep Breeding Research Station, Sandynallah, TamilNadu.

Shahabadi

Shahabadi is found in Shahabad, Patna and Gaya districts of Bihar/Jharkhand state. The animals are of medium size and leggy. The fleece is mostly grey and sometimes with black spots. The ears are of medium size and drooping. The tail is extremely long and thin. Both sexes are polled. Their legs and belly are devoid of wool. Fleece is extremely coarse, hairy and open. Average six monthly greasy fleece yield is 240 g. Fiber diameter is 49.83 µ with Medullation percent of 87.08.

Sonadi

Sonadi is found in Udaipur and Dungarpur districts and, to some extent, Chittorgarh district of Rajasthan and also extends to northern Gujarat. The Animals are fairly well built, somewhat smaller than Malpura, with long legs. Light brown face with the colour extending to the middle of the neck. Ears are large, flat and drooping; Ears generally have a cartilaginous appendage. Tail is long and thin. The fleece is white, extremely coarse and hairy. Belly and legs are devoid of wool. The overall least square means for 1st six monthly and adult six monthly were 528 and 417g respectively and least square means of birth, 3, 6, 9 and 12 month's weight of lambs were 2.52, 10.78, 14.98, 17.32 and 21.76 kg, respectively under farm conditions.

Tibetan

The Tibetan breed is found in northern Sikkirn and Kameng district of Arunachal Pradesh. Animals are medium sized, mostly white with black or brown face; brown and white spots are also found on the body. Both the sexes are horned. The nose is convex, giving a typical Roman nose. The ears are small, broad and drooping. The fleece is relatively fine and dense. The belly, legs and faces are devoid of wool. Average six monthly greasy fleece yield ranges between 400-900 g. Staple length is 7.24 cm, Fiber diameter is 19.3 µ with Medullation percent of 13.22.

Tiruchy Black

Distributed in Perambalur, Ariyalur and Tiruchy, Kallakurichy taluk of South- Arcot district, Viraganur area of Attur taluk of Salem district, Tirupathur and Tiruvannamalai taluks of North Arcot district of TamilNadu. Males are horned and ewes are polled. Ears are short and directed downward and forward. Tail is short and thin. The fleece is extremely coarse, hairy and open. Average body weights at birth, three, six and twelve months of age were 2.13, 9.46, 10.73 and 16.80 Kg, respectively.

Vembur

Distributed in Vembur, Melakkarandhai, Keezhakarandhai, Nagalpuram, Kavundhanapatty, Achangulam and some other villages of Pudur Panchayat Union and Vilathikulam Panchayat areas of Tirunelveli district of Tamil Nadu. Tall animals colour is white, with irregular red and fawn patches all over the body. Ears are medium-sized and drooping. Tail is thin and short. Males are horned and ewes are polled. The body is covered with short hairs which are not shorn. Average body weights at birth, three, six and twelve months of age were 1.97, 8.42, 10.50 and 16.50 Kg, respectively.

Exotic Sheep Breeds

Mutton type	Dorset Horn, Suffolk, Cheviot, Southdown
Wool type	Merino, Rambouillet
Fur type	Karakul

- Merino
- Rambouillet

- Corriedale
- Dorset
- Suffolk
- Karakul

Merino

This breed, known from as early as the twelfth century, was developed in Spain and is believed to have been originally brought from Phoenicia and Carthage with possibly some African sheep inheritance also mixed. The characteristic of this breed is the fine wool fleece, beautifully crimped. The inner fleece is kept soft and pliable by a coating of wool oil on the surface which forms a dark protective covering with dust. Merinos were imported into America in 1801. It has spread very widely in the semi-arid climates of the United States, USSR, South America, Argentina, France and Germany. Rams weigh about 60-80 kg and ewes about 45-60 kg. Good ewes shear 7 to 9 kg of wool1 and rams 10 to 14 kg of wool in a year. This breed is medium sized with a short-head. Ewes have no horns but rams have horns, turning towards the head.

Ramboulliet

This breed was formerly called French Merino. This originated at Rambouillet in France from 1785 by importations of Merino sheep from Spain. These were developed into larger sheep with heavier fleece. This breed was imported into America in 1840. Modern Rambouillet is a large rugged breed with wide heads and well balanced horns curving backward and outward in rams. Rams weigh about 100-140kg and ewes about 70-100kg.

The animals have good conformation for mutton. Average annual yield of fleece is about 5 kg of long fleece with a clean wool percentage of about forty.

Corridale

The Corriedale was developed in New Zealand and Australia during the late 1800s' from crossing Lincoln or Leicester rams with Merino females. The breed is now distributed worldwide, making up the greatest population of all sheep in South America and thrives throughout Asia, North America and South Africa. Its popularity now suggests it is the second most significant breed in the world after

Merinos. The Corriedale is a dual-purpose sheep. It is large-framed, polled with good carcass quality. The Corriedale produces bulky, high-yielding wool ranging from 31.5 to 24.5 micron fiber diameter. The fleece from mature ewes weigh from 4.5-7.7 kg with a staple length of 9-15 cm. The yield percent of the fleece ranges from 50 to 60 percent. Mature rams will weigh from 80-125 kg, ewe weights range from 60-80 kg. The breed is found in most sheep areas of Australia, but mainly in the temperate, higher rainfall zones supporting improved pastures.

Dorset

The Merino sheep were brought into Southwest England and were crossed with the Horned Sheep of Wales, which produced a desirable all-purpose sheep which met the needs of that time. Thus began a breed of sheep which spread over Dorset, Somerset, Devon, and most of Wales and were called Dorset (two varieties namely horned and polled). Dorset is a white sheep of medium size having good body length and muscle conformation to produce a desirable carcass. The fleece is very white, strong, close and free from dark fiber. Dorset fleeces average 2.25-4 kg in the ewes with a yield of between 50% and 70%. The staple length ranges from 6-10 cm with a numeric count of 46's-58. The fiber diameter will range from 33.0 to 27.0 microns. Dorset ewes weigh from 70 to 90 kg at maturity; some in show condition may very well exceed this weight. Dorset rams weigh from 100 to 125 kg at maturity. The ewes are good mothers; good milkers and multiple births are not uncommon.

Suffolk

The Suffolk is the result of crossing Southdown rams on Norfolk Horned ewes. Today's Suffolk derives its meatiness and quality of wool from the old original British Southdown.

The Norfolk Horned sheep, now rare, were a wild and hardy breed. Mature weights for Suffolk rams range from 110-160 kg, ewe weights vary from 80-110 kg. Fleece weights from mature ewe are between 2.25-3.6 kg with a yield of 50 to 60 percent. The fleeces are considered medium wool type with a fiber diameter of 25.5 to 33.0 microns and a spinning count of 48 to 58. The staple length of Suffolk fleece ranges from 5-8.75 cm.

Karakul

This is a fur bearing breed of sheep, native to Bokhara, situated between Pakistan and Afghanistan in central Asia. The name is derived from the area Karakul (black lake) which is very un-favourable for raising livestock. This breed originated by crossing the two different breeds of this area. This was imported into the United States in 1909. The modern Karakul is a medium-sized breed, rams weighing about 80-90 kg and ewes about 60-70 kg. This produces light white fleece of low-grade averaging about 3 kg per year. The breed is poor in mutton quality. The chief virtue of this breed is the fur pelt of the baby lambs. The pelt is taken from a prematurely born lamb or Iamb killed within a short time after birth.

Synthetic Sheep Breeds Evolved in India

Avikalin

Evolved at Central Sheep and Wool Research Institute, Aviaka Nagar, Rajasthan for carpet wool by crossing Rambouillet with Malpura. The exotic inheritance is stabilised around 50 per cent.

Avivastra

Evolved at, Central Sheep and Wool Research Institute, Aviaka Nagar, Rajasthan for fine wool by crossing Rambouillet with Chokla. The exotic inheritance is stabilised around 50-75 per cent.

Avimaans

Evolved at Central Sheep and Wool Research Institute, Aviaka Nagar, Rajasthan for mutton by crossing Dorset / Suffolk rams with Malpura/Sonadi ewes. The exotic inheritance is stabilised around 50-75 per cent.

Bharat Merino

Evolved at Central Sheep and Wool Research Institute, Aviaka Nagar, Rajasthan for fine wool by crossing Soviet Merino rams with Chokla / Nali ewes. The exotic inheritance is stabilised around 50-75 per cent.

Hissardale

Evolved at Hissar for fine wool by crossing Australian Merino rams with Bikaneri ewes. The exotic inheritance is stabilised around 50-75 per cent.

Kashmir Merino

Evolved from crosses of different Merino types with native ewes. The exotic inheritance ranges from 50-75 per cent.

Nilagiri Synthetic (Sandyno)

Evolved at Sheep Breeding Research Station, Sandynallah, The Nilgiris, Tamilnadu for fine wool by crossing Russian Merino rams and Rambouillet rams with Nilagiri ewes. The exotic inheritance is stabilised around 50 per cent.

Nellore Synthetic

Evolved at Livestock Farm, Palamaner, Andhra Pradesh for mutton by crossing Dorset rams with Nellore ewes. The exotic inheritance is stabilised around 50 per cent.

Classification of Indian Goat Breeds

Goat (domestic group)	
Phylum	Cordata
Sub-phylum	Craniata (Vertebrata)
Class	Mammalia
Sub-class	Theria (Viviparous)
Infra-class	Eutheria (Placenta)
Order	Ungulata (hoofed mammals)
Sub-order	Artiodactyles (even-toed)
Sub-division	Pecora (true ruminants)
Family	Bovidae (hollow -horned)
Genus	Capra
Species	hircus

Other Groups of Goats

True group, Pasang group, Ibex group and Markhor group

Northwestern arid and semi arid region	Northern temperate Region	Southern peninsular region	Eastern region
Sirohi	Gaddi	Sangamneri	Ganjam
Marwari	Changthangi	Malabari	Bengal
Beetal	Chigu	Osmanabadi	
Jhakarna		Kanni adu	
Barbari			
Jamnapari			
Mehsana			
Gohilwadi			
Kutchi			
Surti			

Barbari

Distribution

District of Rajasthan, Etah, Agra and Aligarh districts of Uttar Pradesh, and Bharatpur. The total goat population in the Barbari distribution area, according to the 1972 census, was 0,444 m, of which 0,028 m adult males and 0,270 m adult females. Officials of the State Animal Husbandry Department state that Barbari goats true to breed number only about 30000. The breed, quite important for milk and meat, is a dwarf breed highly suited for rearing under restrained and stall-feeding conditions. In addition to being a good milkier, it is highly prolific. Considering the number of animals reported, there is serious need for undertaking conservation measures. It was observed in personal surveys that most males are castrated early in life and fattened for slaughter at religious festivals, and that a sizeable number of non-descript goats are kept for milk production by owners who maintain Barbaris, as all the milk of the latter is allowed to be suckled by their kids to ensure good growth.

Breed characteristics

	Adult males	Adult females
Body weight	37.85	22.56
Body length	70.45	58.68
Height at withers	70.67	56.18
Chest girth	75.53	64.31

Conformation

Small animals, with compact body. The orbital bone is quite prominent, so that eyes appear bulging. There is wide variation in coat colour, but white with small light brown patches, is the most typical. Ears are short, tubular, almost double) with the slit opening in front, erect, directed upward and outward. Both sexes have twisted horns, medium in length and directed upward and backward; horn length: 11.17 cm. Bucks have a large thick beard.

Reproduction

	Days
Age at first kidding	648
Kidding interval	348
Service period	70
Kidding percentage (%)	70.2
Litter size: (%)	
Singles	49.64
Twins	49.32
Triplets	1.04
Quadruplets	

Performance

Milk

Lactation yield (Kg)	107.120
Lactation length (days)	150.13

Meat (kg.)

At birth	1.739
At weaning	6.661
6 months	7.800
9 months	12.566
12 months	14.517

Bengal

Individuals of this breed are sometimes distinguished by color as Black, Brown, Gray and White Bengal. Personal surveys suggest that the breed could be separated into two types, one found in the hot humid plains of West Bengal, Bihar and Orissa, and the other in the sub-temperate and humid uplands and plains of Assam and other northeastern States. The latter were earlier known as Assam Plain and Assam Hill goats, depending on their location but are not substantially different from the former.

Distribution

Distributed throughout all eastern and northeastern India, from Bihar through northern Orissa to all West Bengal, Assam, Manipur, Tripura, Arunachal Pradesh and Meghalaya. The total Bengal-type goat population in the States of West Bengal, Bihar and Orissa, according to the 1972 census, was 14.164 m, of which 2.198 m adult males and 6.257 m adult females. In Assam and other northeastern States, the goat population was 1.501 m, of which 0.304 m adult males and 0.543 m adult females.

Breed characteristics

	Adult males	Adult females
Body weight		20.38
Body length	63.2	51.2
Height at withers	58.3	55.4
Chest girth	72.0	63.2

Assam and other North-estern region

	Adult males	Adult females
Body weight	15.38	14.27
Body length	53.8	54.8
Height at withers	49.5	49.3
Chest girth	58.7	59.1

Conformation

Small animals. The predominant coat colour is black, brown, gray and white are also found, the former two sometimes with black markings along the back and on the belly and extremities. In a survey of 200goats in Assam, 96 were completely black, 1 completely white, 11 brown, 45 white with black markings and 17 brown with black markings. In another survey in Bihar, the proportion of gray to brown as 35:65. The hair coat is short and lustrous. The nose line is slightly depressed. Both sexes have small to medium horns, directed upward and sometimes backward, average horn length 6.4 cm (range : 5.8 to 11.5 cm). beared is observed in both sexes. The ears are short, flat and carried horizontally, average ar length 13.8 cm (range:11.5 to 4.1 cm)

Reproduction

Kidding percentage (%)		
	Singles	44.6
	Twins	51.3
	Triplets	4.1

Performance

Milk		
	Lactation yield (Kg)	58
	Lactation length (days)	118

Meat (kg.)		
	At birth	1.31
	At weaning	6.09
	6 months	8.80
	9 months	12.60

Chigu

Distribution

Lahaul and Spiti valleys of Himachal Pradesh, arid Uttar Kashi, Chamoli, Pithoragarh districts of Uttar Pradesh, bordering Tibet. Mountainous ranges with the altitude varying from 3500 to 5000 m. The area is mostly cold and arid.

Breed characteristics

	Adult males	Adult females
Body weight	39.42	25.71
Body length	75.8	69.3
Height at withers	68.6	60.0
Chest girth	80.70	73.3

Conformation

Medium-sized animals. The coat is usually white, mixed with grayish red. Both sexes have horns, directed upward, backward and outward, with one or more twists. These goats are not very different in conformation from Changthangi.

Reproduction

	Days	
	Age at first kidding	615.8
	Kidding interval	272.8
	Service period	
	Kidding percentage (%)	65.4
	Litter size: (%)	
	Singles	99.2
	Twins	0.8

Performance

Meat (kg.)		
	At birth	2.10
	At weaning	8.41
	6 months	12.17
	9 months	14.75
	12 months	18.46
	Hair	
	Average fiber length (cm)	5.9
	Average production (g)	120.31
	Average fiber diameter (µ)	11.77

Gaddi

Distribution : Chamba, Kangra, Kulu, Bilaspur, Simla, Kinnaur and lahaul and Spiti in Himachal Pradesh and Dehradun, Nainital, Tehrigarhwal and Chamoli hill districts in Uttar Pradesh. The total goat population in the Gaddi distribution area, according to the 1972 census, was 0.770 m, of which 0.125 m adult males and 0.468 m adult females.

Breed characteristics

	Adult males	Adult females
Body weight	27.45	24.72
Body length	69.5	65.2
Height at withers	61.3	58.1
Chest girth	72.2	69.3

Conformation

Medium-sized animals. Goat colour is mostly white, but black and brown and combinations of these are also seen. Both sexes have large horns, directed upward and backward and occasionally twisted. Ears are medium long and drooping. The nose line is convex. The udder is small and rounded, with small teats placed laterally. The hair is white, lustrous and long. Flock size ranges from 20 to 500.

Reproduction: Essentially single, twinning occurs in only 15 to 20% of births.

Performance

Meat (kg.)	380
Hair	
Average fleece yield per clip (g)	300
Medullation (%)	74.48
Average fiber diameter (μ)	73.4

Ganjam

Distribution

Southern districts of Orissa: Ganjam and Koraput. The total goat population in the Ganjam distribution area, according to the 1972 census, was 0.448 m, of which 0.056 m adult males and 0.103 m adult females.

Breed characteristics

	Adult males	Adult females
Body weight	44.05	31.87
Body length	76.2	67.6
Height at withers	84.5	77.1
Chest girth	83.1	74.6

Conformation

Tall, leggy animals. The coat may be black, white, brown or spotted, but black predominates. Hairs are short and lustrous. Ears are medium sized; ear length: 14.50 +- 0.15 cm. Both sexes have long, straight horns, directed upward; horn length: 20.9+- 0.33 cm. Tail is medium-long. The average flock contains 84.5 +- 37.5 individuals range: 20 to 500, of which 2.9 adult males, 59.8 adult females and 21.8 young.

Reproduction

	Days	
Kiddinginterval		376
Serviceperiod		218.6
Kidding percentage (%)		82
Litter size (%)		
Singles		98.4
Twins		1.6

Performance

Milk		
	Lactation yield (Kg)	319.44
	Lactation length (days)	141.6
Meat (kg.)		
	At birth	2.31
	At weaning	9.52
	6 months	
	9 months	
	12 months	11.69

Jamnapari

The name is derived from the location of the breed beyond the river Jamna (JamnaPar) in Uttar Pradesh. (A distinct strain.called Ramdhan (also known locally as Kandari Ka Khana) originated from a cross between a doe from the Alwar region in Rajasthan and a Jamnapari buck; it has a typical nose shape, even more like a parrot mouth than is usually observed in the true breed. The strain is known to have better milk production and growth, but its flocks are very limited in number.

Distribution

Agra, Mathura and Etawa districts in Uttar Pradesh and Bhind and Morena districts in Madhya Pradesh. However the pure stocks are found only in about 80 villages in the vicinity of Batpura and ChakarNagar in Etawa district. The total goat population in the Jamnapari distribution area, according to the 1972 census, was 0.58 m. However, officials of the Animal Husbandry Department of Uttar Pradesh state that the total number of pure-bred Jamnapari does not exceed 5000; these are located mostly in the ChakarNagar area, between the Jamna and Chambal ravines. There is a serious need for conservation, multiplication and further improvement of the breed, considering the extremely small numbers of pure-bred animals remaining. Jamnapari is one of the largest goats in India; it has been extensively utilized to upgrade indigenous breeds for meat and milk, and has been taken to neighboring countries for the same purpose.

Breed characteristics

	Adult males	Adult females
Body weight	44.66	38.03
Body length	77.37	75.15
Height at withers	78.17	75.20
Chest girth	79.52	76.11

Conformation

Large animals. There is a great variation in coat color, but the typical coat is white with small tan patches on head and neck. The typical character of the breed is a highly convex nose line with a tuft of hair, yielding a parrot mouth appearance. The ears are very long, flat and drooping; ear length: 26.79 cm. Both sexes are horned; horn length: 8.69 cm. Tail is thin and short. A thick growth of hair on the buttocks, known as feathers, obscures the udder when observed from behind. The udder is well developed, round, with large conical teats. The average flock contains 16.0 (range 8 to 41) of which 0.25 adult males 8.65 adult females and 7.1 young.

Reproduction

Days	
Age at first kidding	737
Kidding interval	229
Service period	101
Kidding percentage (%)	79.6
Litter size: (%)	
Singles	56.2
Twins	43.1
Triplets	0.7

Breeding: Flocks are pure-bred. Selection in bucks is based on dam's milk.

Performance

Milk	
Lactation yield (Kg)	201.96
Lactation length (days)	191
Meat (kg.)	
At birth	4.27
At weaning	12.11
6 months	15.56
9 months	24.00
12 months	29.65
Hair (g per year)	

Marwari

Distribution

Marwari region of Rajasthan, comprising Jodhpur, Pal, Nagaur, Bikaner, Jalore, Jaisalmer and Barmer districts. The breed also extends into certain areas of Gujarat, especially Mehsana district. The total goat population in the Marwari distribution area, according to the 1972 census, was 3.914 m, of which 0.072 m adult males and 2.484 m adult females.

Breed characteristics

	Adult males	Adult females
Body weight	33.18	25.85
Body length	70.97	63.51
Height at withers	74.74	69.29
Chest girth	71.68	68.60

Conformation

Medium-sized animals. Predominantly black with long shaggy hair coat. In about 5% of individuals, white or brown patches are also observed. Beard is present in both sexes. Ears are flat, medium in length and drooping; ear length: 16.38 + 0.20 cm. Both sexes have short, pointed horns, directed upward and backward; horn length: 10.10 + 0.18 cm. Tail is small and thin. Udder is small and round, with small teats placed laterally. Average flock size: 46.0 to 48.0 (range: from 10 to 100) containing 0.6 adult males, 36.4 adult females and 11.0 young.

Reproduction

Kidding percentage	52.5
Singles	Generally

Breeding: Pure breeding. Breeding males are selected on size and hair production; on, mostly from within the flock. The Animal Husbandry Department of the Government of Rajasthan has introduced cross-breeding with Jamnapari; to increase milk production.

Performance

Milk		
	Lactation yield (kg)	91.39
	Lactation length (days)	105.80
Meat		
	At birth	2.29
	At weaning	6.0
	6 months	8.70
	9 months	13.70
	12 months	16.25
	Hair (g. per year)	302.9

Mehsana

Distribution

Banaskantha, Mehsana, Gandhi Nagar and Ahmedabad districts of Gujarat. The total goat population in the Mehsana distribution area, according to the 1972 census, was 0.736 m, of which 0.030 m adult males and 0.530 m adult females.

Breed characteristics

	Adult males	Adult females
Body weight	37.14	32.39
Body length	71.2	68
Height at withers	80.4	74.3
Chest girth	76.9	73

Conformation

Large animals. The coat is black, with white spots at the base of the ear. Nose line is straight. The hair coat is long and shaggy. Ears are white, leaf-like and drooping; ear length: 15.8 cm. Both sexes have slightly twisted horns, curved upward and backward horn length: 11 cm .The udder is well developed; the teats are large and conical. The average flock contains 54.7 individuals (range: 20 to 300), of which 0.7 adult males, 39.4 adult females and 14.6 young.

Reproduction

	Kidding percentage (%)	69.5
	Litter size: (%)	
	Singles	89.7
	Twins	10.3

Breeding: Pure breeding. Males are selected primarily on body size.

Performance

	Milk	
	Lactation yield(Kg)/day	1.32
	Lactation length (days)	197
	Hair (g per year)	210

Malabari

Distribution

Calicut, Cannannore and Malapuram districts of Kerala. The total goat population in the Malabari distribution area, according to the 1972 census, was 0.389 m, of which 0.028 m adult males and 0.193 m adult females.

Breed characteristics

	Adult males	Adult females
Body weight	38.96	31.12
Body length	70.2	63.5
Height at withers	71.9	63.2
Chest girth	73.8	67.4

Conformation

Medium-sized animals. Coat color varies widely from completely white to completely black. 31% of the goats have long hair. Males and a small percentage of females (13%) are bearded. Both sexes have small, slightly twisted, horns, directed outward and upward. Ears are medium-sized, directed outward and downward; ear length: 16.20 +- 0.17 cm. Tail is small and thin; tail length: 13.16 +- 0.06 cm. Udder is small and round, with medium-sized teats. The average flock contains 5.44 individuals (range: 1 to 15), of which 0.40 bucks, 2.02 does and 3.02 kids. Two thirds of flocks do not maintain a breeding male.

Reproduction

	Days	
Age at first kidding		609.9
Kidding interval		286.6
Litter size: (%)		
Singles		50.5
Twins		42.4
Triplets		6.6
Quadruplets		0.5

Breeding

Pure breeding except in very limited areas near Mannuthy, where cross-breeding with Alpine and Saanen has been undertaken by the Kerala Agricultural University. The cross-breeds showed improvement both in reproduction and milk production.

Performance

	Milk	
	Lactation yield (kg)	43.78
	Lactation length (days)	143.5
	Meat (Kg)	
	At birth	1.63
	At weaning	5.76
	6 months	8.73
	9 months	11.41
	12 months	14.12

Kutchi

Distribution

Kutch district in Gujarat. The total goat population in the Kutch district, according to the 1972 census was 0.402 m. of which 0.018 m adult males and 0.298 m adult females.

Breed characteristics

	Adult males	Adult females
Body weight	43.5	39.29
Body length	77.1	75.0
Height at withers	86.4	82.4
Chest girth	78.4	76.1

Conformation

Large animals. The coat is predominantly black but a few white, brown and spotted animals are also found. The hair is coarse and long. The nose is slightly Roman. The ears are long, broad and drooping; ear length: 22.0 ± 0.26 cm. Both sexes have short, thick horns, pointed upward; horn length: 10 ± 0.19 cm. The udder is reasonably well developed; teats are conical.

Flock structure: The average flock contains 83.3 ± 14.5 individuals (range: 30 to 300), of which 2.4 adult males, 70.3 adult females and 10.6 young.

Reproduction

	Days
Kidding percentage (%)	75.5
Litter size: (%)	
Singles	84.1
Twins	15.2
Triplets	0.7

Performance

	Milk
Lactation yield (kg)	1.84
Lactation length (days)	117.7
Hair (g per shorn)	229.3 twice a year

Kannaiadu

Distribution

Ramnathapuram and Tirunelveli districts in Tamil Nadu. Numbers: The total goat population in the Kannaiadu distribution area, according to the 1972 census, was 0.604 m, of which 0.093 m adult males and 0.295 m adult females.

Breed characteristics

	Adult males	Adult females
Body weight	35.76	28.62
Body length	71.06	67.30
Height at withers	84.12	76.15
Chest girth	77.53	70.83

Conformation

Tall animals, predominantly black or black with white spots. Ears are medium-long; ear length: 15.67+- 0.13 cm. Males are horned; females are polled. Tail is medium-sized and thin; tail length: 15.77 +- 0.47 cm. Udder is small and round, with small teats placed laterally. The average flock contains 17 individuals (range: 2 to 40 (61), of which 1 buck, 11 does and 6 young.

Reproduction

Kidding percentage (%)	80-85
Litter size: (%)	
Singles	90
Twins	10

Performance: Animals are maintained for meat purposes and are not milked.

Jhakrana

Distribution

Jhakrana and a few surrounding villages near Behror, in the Alwar district of Rajasthan. The number of animals of this breed is rather small. As it is restricted to a very limited area. Considering that it is a good indigenous dairy breed, there is need for its conservation.

Breed characteristics

	Adult males	Adult females
Body weight	57.8	44.48
Body length	84.10	77.74
Height at withers	90.40	79.12
Chest girth	86.00	79.13

Conformation

Large animals. The coat, predominantly black with white spots on ears and muzzle, is short and lustrous. Face line is straight. Forehead is narrow and slightly bulging. The breed is quite similar to Beetal, the major difference being that Jhakrana is longer. Ear length is medium: 13.8 + 0.19. Udder is Plarge, with large conical teats. The average flock consists of 14.2+- 2.1 individuals; (range: 5 to 34), including 0.05 adult males, 9.08 adult females and 5.07 young.

Reproduction

Litter size: (%)	
Singles	57
Twins	41
Triplets	2

Breeding

Pure breeding. Males are selected on the basis of their dam's milk yield. Breeding bucks are generally produced from within the flock. Owners of small flocks who do not maintain a buck utilize services of the buck from a neighbor's flock.

Performance

Milk	
Lactation yield (kg)	121.80
Lactation length (days)	114.7

Sirohi

Distribution

Sirohi district of Rajasthan. The breed also extends to Palanpur in Gujarat. The total goat population in the Sirohi; distribution area, according to the 1972 census, was 0.295 m, of which 0.007 m adult males and 0.204 m adult females.

Breed characteristics

	Adult males	Adult females
Body weight	50.37	22.54
Body length	80.0	61.3
Height at withers	85.6	68.4
Chest girth	80.3	62.4

Conformation

Compact, medium-sized animals. Coat color predominantly brown, with lignt or dark brown patches; a very few individuals are completely white. Most animals are wattled. Ears are flat and leaf-like, medium-sized and drooping; ear length: 18.8 + _0.6 cm (15). Both sexes have small horns curved upward and backward; horn length: 7.7 ~ 0.15 cm (144). Tail is medium in length and curved upward; tail length: 16.7 ~ 0.14 cm (153). Udder is small and round with small teats placed laterally. Average flock size is 60 (range: 10 to 200) containing 1 adult male, 42 adult females and 17 young.

Reproduction

	Kidding interval	89.3%
	Litter size: (%)	
	Singles	91.5%
	Twins	8.5%

Breeding: Generally pure breeding. Males are selected on size from within flocks. There is some introduction of Marwari for increasing hair production.

Performance

	Milk	
	Lactation yield (kg)	71.18
	Lactation length (days)	174.8
	Meat (Kg)	
	At birth	2.82
	At weaning	9.92
	6 months	13.48
	9 months	16.95
	12 months	21.27

Sangamneri

Distribution

Poona and Ahmednagar districts of Maharashtra. The total goat population in the Sangamneri distribution area, according to the 1972 census, was 5.692 m, of which 0.396 m adult males and 3.439 m adult females.

Breed characteristics

	Adult males	Adult females
Body weight	38.37	28.97
Body length	69.8	62.5
Height at withers	77.3	68.0
Chest girth	76.0	71.0

Conformation

Medium-sized animals. Body color may be white, black or brown, with spots of the other colors. Ears are medium-sized and drooping; ear length: 15.89+- 0.62 cm. Both sexes have horns, directed backward and upward length: 12.36 +- 0.58 cm. Tail is thin and short; tail length: 15.72 + -0.32 cm (33). The hair coat is extremely coarse and short. The average flock contains 30.5+- 12.63 individuals (range: 6 to 91), of which 1.0 adult male, 14.5 adult females and 15.0 young.

Reproduction

	Days	
Age at first kidding	422.2	
Kidding interval	333.7	
Service period	155.6	
Litter size: (%)		
Singles	69.9	
Twins	30.0	
Triplets	0.5	

Breeding : Pure breeding. Males are selected on dam's milk yield.

Performance

	Milk	
Lactation yield (kg)	83.4	
Lactation length (days)	165	
Meat (Kg)		
At birth	1.86	
At weaning	7.09	
6 months	10.06	
9 months	13.44	
12 months	17.33	
Hair (g per clip)	250-300	

Osmanabadi

Distribution

Latur Tuljapur and Udgir taluks of Osmanabadi district of Maharashtra. The total goat population in the Osmanabadi distribution area was 0.219 m. of which 0.020 m. adult males and 0.119 m. adult females.

Breed characteristics

	Adult males	Adult females
Body weight	33.6	32.36
Body length	69.12	67.51
Height at withers	77.87	74.79
Chest girth	72.06	72.04

Conformation

Animals surveyed 73% were black and the rest were white, brown or spotted. Ears are medium long; ear length: 18.0 +- 0.10 cm. Most males (89.5%) are horned; females may be horned or polled, in almost equal proportions. Tail is medium long and thin; tail length: 16.6 +- 0.10 cm. The udder is small and round with small teats placed laterally. The average flock contains 10.73 individuals (range:1 to 65) of which 0.28 adult males; 6.56 adult females and 3.89 young.

Reproduction

Kidding percentage (%)	129.9
Litter size: (%)	
Singles	70.5
Twins	29.0
Triplets	0.5
Quadruplets	

Breeding : Pure breeding. There is little selection.

Performance

	Milk	
	Lactation yield (kg)	0.5-1.5
	Lactation length (days)	3-5
	Meat (Kg)	
	At birth	2.39
	At weaning	7.34
	6 months	11.07
	9 months	15.12

Zalawadi

Distribution

Surendranagar and Rajkot districts in Gujarat. The total goat population in the Zalawadi distribution area according to the 1972 census was 0.341 m. of which 0.013 m adult males and 0.256 m adult females.

Breed characteristics

	Adult males	Adult females
Body weight	38.84	32.99
Body length	75.6	71.8
Height at withers	83.3	78.5
Chest girth	76.8	74.2

Conformation

Large animals. Coat is black and contains long coarse hair. Ears are long, wide, leaf-like and drooping; ear length: 19.0 +- 0.20 cm. Both sexes have long twisted horns, pointed upward; horn length: 1673 +- 0.2 cm. The udder is well developed, with large conical teats. The average flock contains 41.5 individuals (range: 10 to 60), of which 0.9 adult males, 28.7 adult females and 11.9 young.

Reproduction

	Days	
	Kidding interval	71.2
	Litter size: (%)	
	Singles	82.1
	Twins	17.9

Performance

	Milk	
	Lactation yield (kg)	2.02
	Lactation length (days)	197.2
	Hair (g per year)	245.3

Beetal

Distribution

Throughout the States of Punjab and Haryana. True-bred animals are however found in the districts of Gurdaspur, Amritsar and Ferozepur in Punjab. The total goat population in the Beetal distribution area, according to the 1972 livestock census, was 0.159 m, of which 0.014 m adult males and 0.084 m adult females. The goat population of Punjab has shown a marked decline. Surveys carried out by the Division of Dairy Cattle Genetics of the National Dairy Research Institute (NDRI) showed that both the number of flocks and the number of goats per flock have decreased, primarily owing to an increase in irrigated cultivated areas and the shortage of natural vegetation available for browsing. Beetal is a good dairy breed, second to Jamnapari in size but is superior to it in that it is more prolific and more easily adaptable to different agro-ecological conditions and to stall-feeding. Jamnapari is more sensitive.

Characteristics

	Adult males	Adult females
Body weight	59.07	34.97
Body length	85.7	0.42
Height at withers	91.60	77.13
Chest girth	86.0	73.7

Conformation

Large animals. Variable coat color, predominantly black or brown with white spots of differing sizes. In a survey conducted by NDRI in the home tract of the breed, 92.6% animals were black and 7.4% brown. The coat is short and lustrous. The face line is convex, with typical Roman nose but not as prominent as in Jamnapari. Ears are long and flat, curled and drooping ear length: 24.8 t 0.65 cm.

Both sexes have thick, medium-sized horns, carried horizontally with a slight twist directed backward and upward; horn length: 11.95 t 0.76 cm. Tail is small and thin. The udder is large and well developed, with large conical teats. The average flock contains 21.06 + 1.92 individuals, of which 1.5 adult males, 11.7 adult females and 7.8 young.

Reproduction

	Days	
	Age at first kidding	761
	Kidding interval	368
	Service period	160
	Kidding percentage	176
	Litter size: (%)	
	Singles	40.66
	Twins	52.6
	Triplets	6.52
	Quadruplets	0.22

Breeding

Pure breeding. Breeding males are generally selected on the basis of their dam's milk yield. The services of a buck of a neighboring owner are commonly utilized when a flock does not possess a buck.

Performance

	Milk	
	Lactation yield (kg)	177.38
	Lactation length (days)	187.0
	Meat (Kg)	
	At birth	2.80
	At weaning	9.26
	6 months	12.18
	9 months	15.42
	12 months	21.83

Chanthangi

Distribution

Changthang region of Ladakh, at altitudes above 4000 m. The goat population in this region, was approximately 0.04 m.

Climate

A cold arid region. Average annual precipitation: 9.26 cm, distributed throughout the year, with maximal during January/April. Summer and winter temperatures are extreme (+40 ° C to -40 ° C). Most cultivation takes place along the rivers.

Breed Characteristics

	Adult males	Adult females
Body weight	20.37	19.75
Body length	49.8	52.4
Height at withers	49.0	51.6
Chest girth	63.0	65.2

Conformation

Medium-sized animals. Half of the animals are white, the remainder black, gray or brown. Both sexes have horns, generally large (range: 15 to 55 cm), turning outward, upward and inward to form a semi-circle, but a wide variation exists in both shape and size. The flock size ranges from 200 to 300 in flocks belonging to migratory shepherds. In the stationary flocks, it is between 10 and 15.

Reproduction

In farmers' flocks: kidding percentage: 65%. Under farm conditions: kidding percentage: from 80 to 90%.

Performance

Meat (Kg)		
	At birth	2.18
	3 months	7.76
	9 months	9.18
	12 months	11.80
	Hair	
	Average fiber length (cm)	4.94
	Scouring yield (%)	65.28
	Average fiber diameter (μ)	13.86

Exotic Goat Breeds

Milch breed	Saanen Alpine
Mohair breed	Angora
Dual purpose	Anglo-nubian

Saanen

The Saanen dairy goat is originated in Switzerland in the Saanen Valley. Saanen does are heavy milk producers and usually yield 3-4 percent milk fat. It is medium to large in size, weighing approximately 65kg with rugged bone and plenty of vigor. Saanens are white or light cream in color. Ears should be erect and alertly carried, preferably pointing forward. The face should be straight or dished. The breed is sensitive to excessive sunlight and performs best in cooler conditions. The provision of shade is essential and tan skin is preferable.

Toggenburg

The Toggenburg is a Swiss dairy goat from Toggenburg Valley of Switzerland. They are also credited as being the oldest known dairy goat breed. This breed is medium size, sturdy, vigorous, and alert in appearance. Slightly smaller than the other Alpine breeds. Does weight is 55kg. Colour is solid varying from light fawn to dark chocolate with no preference for any shade. Distinct white markings and varying degrees of cream markings are acceptable. The ears are erect and carried forward. Facial lines may be dished or straight, never roman. Toggenburgs perform best in cooler conditions. They are noted for their excellent udder development and high milk production, and have an average fat test of 3.7 percent.

Alpine

Alpine is a breed of goat that originated in the Alps. There are varieties of Alpine namely French Alpine and British Alpine. No distinct colour has been established, and it may range from pure white through shades of fawn, grey, brown, black, red, bluff, piebald, or various shadings or combinations of these colours. Both sexes are generally short haired, but bucks usually have a roach of long hair along the spine. The beard of males is also quite pronounced. The ears in the Alpine should be of medium size, fine textured, and preferably erect. Mature females weigh not less than 60 kg. Males weigh not less than 75 kg. Alpine females are excellent milkers and usually have large, well-shaped udders with well-placed teats of desirable shape. These are hardy, adaptable animals that thrive in any climate while maintaining good health and excellent production.

Angora

The Angora goat originated in the district of Angora in Asia Minor. Angora stock was distributed to different countries. The most valuable characteristic of the Angora as compared to other goats is the value of the mohair that is clipped. The average goat shears approximately 2.4 kg of mohair per shearing and are usually sheared twice a year. They produce a fiber with a staple length of between 12 and 15cm. The mohair is very similar to wool in chemical composition but differs from wool in that it is has a much smoother surface and very thin, smooth scale. Consequently, mohair lacks the felting properties of wool. Mohair is very similar to coarse wool in the size of fiber. It is a strong fiber that is elastic, has considerable luster, and takes dye very well. Mohair has been considered very valuable as an upholstering material for the making of covering materials where strength, beauty, and durability are desired. The market valuation of mohair fluctuates more than does that of wool, but, in general, satisfactory prices are obtained for the clip. During depressed times, the market has favored fine hair and because fine hair is normally shorn from young goats, selection for fertility has also become increasingly important. Both sexes are horned. The bucks usually have a pronounced spiral horns, whereas, the horn of the female is comparatively short, much smaller, and has only a very slight tendency to spiral. The Angora goat is a small animal as compared to sheep, common goats, or milk goats. Mature bucks weigh from 80 to 100 kg but do not reach their maximum weight until after five years of age. Does weight ranges from 30 to 50 kg. The Angora goat is not as prolific as other goats and twins are not the usual birth.

Boer

The Boer is an improved indigenous South African breed with some infusion of European, Angora and Indian goat breeding many years ago. The Boer goat is primarily a meat goat. It is a horned breed with lop ears and showing a variety of color patterns. The Boer goat is being used very effectively in South Africa in combination with cattle due to its browsing ability and limited impact on the grass cover. The mature Boer male weighs between 110-135 kg and Boer females between 90 and 100 kg. Performance records for this breed indicate exceptional individuals are capable of average daily gains over 200 g/day in feedlot. More standard performance would be 150-170 g/day. A kidding rate of 200% is common for this breed. Puberty is reached early, ususally about 6 months for the males and 10-12 months for the females.

Classification of Pigs

Pig Breeds

Indigenous pigs are known as Desi pigs, mostly black in colour and smaller in size when compared to western breeds.

Exotic breeds of pigs are :-

- Large white Yorkshire
- Middle White Yorkshire
- Landrace
- Berkshire
- Hampshire
- Tamworth

Pigs are estimated to have been domesticated in the Neolithic period during B. C. 2500-2400. The domestic breeds were developed from the wild forms in Europe, India and East Asia independently.

Large White Yorkshire

This breed originated in Yorkshire in England. It has been classified into three distinct types - large, medium and small. They have been developed from Licester pigs of Robert Bakewell. These animals moved into the United States in 1893. This is a good bacon breed which has been recently developed for lean meat. They are white in colour with small ears tilting to the front. They have broad face with a medium dish.

Berkshire

This breed has been developed in Berkshire County in South-central England from crosses of old English hog with sows of Chinese and Siamese origin. The English hog was a descendent of the European wild boar Sus scrofa. Pigs of this breed are large, long bodled, heavy-boned animals which stand on long legs and have arched narrow back they weigh about 500 kg. They are very famous for lean meat. As breed, they were established in 1816. They were imported into United States as early as 1823. They have black coat with white feet and white stripes on the face. They have a dished face.

Landrace

This was developed in Denmark about 1895, by crossing English pigs on native pigs especially for the bacon industry. The Landrace has some Chinese pig inheritance. They are white in colour and have a long narrow body.

Duroc

This breed originated in the eastern United States. This is also a breed developed out of crossing pigs imported from Africa, Spain and Portugal. They were formally called Duroc Jersey. Subsequently, some Tamworth inheritance also entered into crossing. They are light golden to dark red in colour. The body is or medium length with tall legs. Adult boars weigh about 450 kg and sow about 350 kg.

Classification of Pigs

Pig	
Phylum	Cordata
Sub-phylum	Craniata (Vertebrata)
Class	Mammalia
Sub-class	Theria (Viviparous)
Infra-class	Eutheria (Placenta)
Order	Ungulata (hoofed mammals)
Sub-order	Artiodactyles (even-toed)
Sub-division	Sunia
Family	Suidae (True pigs)
Genus	Sus
Species	scrofa (European wild pigs) domesticus (domestic pigs) cristatus (Indian wild pigs) and amanensis (Andaman Islands) verrucosus (Malayan pigs)
	vittatus (Malayan pigs)
	barbatus (Malayan pigs) salvanius (Himalayan pigs) africanus, procus (African pigs)

Classification of Fowls

Fowls are often classified based on the purpose for which they are developed such as egg type, meat type and dual purpose (for both egg and meat), but it is mostly on the basis of their origin. According to the latter, the birds are classified into the following major classes: American, Asiatic, English and Mediterranean. A breed refers to a group of domestic fowls with a common ancestry, and having similarity in shape, conformation, growth, temperament, shell colour of egg and breeds true to type. Variety is a subdivision of breed and within a breed there may be several varieties. The term variety is used to distinguish fowls having the characteristics of the breed to which they belong but differing in plumage colour, comb type, etc., from other groups of the same breed. A breed or a variety may have several strains or lines identified by a given name and produced by a breeder through at least 5 generations of closed flock breeding. Several strains within a breed or variety phenotypically may look alike but often differ in their production performance depending upon their breeding history.

Class	Breed	Varieties
American	Plymouth Rock	Barred Whites, Buff Patridge, Silver pencilled
	Wayandotte	Whites, Silver laced, Patridge
	Rhode Island Red	Rose comb, Single comb
	Rhode Island White	Rose comb
	New Hampshire	Single comb
	Jersey Black Giant	Only single comb
English	Orphington	Buff, Whites, Black
	Sussex	Light, Speckled
	Australop	Black
	Dorking	Black
	Cornish	Whites, Rock
Asiatic	Brahama	Light, Dark
	Long Shan	Black, Whites
	Cochin	Buff, Whites, Black, Patridge
Medetaranian	Leghorn	White, Dark, Light, Buff
	Minorca	White, Black, Buff
	Ancona	Single comb, Rose comb
	Blue Andulusian	Single comb

Contd...

Spain	Spanish Butter cup	
Polish	Polish	
Hamburg	Hamburg	
French	Hounder	
	Crevecocus	
Game	Game	
Oriendae	Sumatra	
	Malaya	
	Cubalaya	
Miscellaneous	Sultan Frizzle	

Common Breeds of Poultry

Modern domestic breeds are considered to be the descendants of the jungle fowl found in India and its neighbouring countries like Sri Lanka, Burma, China, Java, Sumatra and Malaysia. Four species of wild or jungle fowl are known, viz. Gallus gallus (red jungle fowl), Gallus lafayettii (Ceylon jungle fowl), Gallus sonneratii (grey jungle fowl) and Gallus varius (Javan jungle fowl). The relative contribution of these few species to the formation of modem domestic breeds still remains controversial. While some believe that all the present-day domestic breeds of poultry have originated from red jungle fowl (Gallus gallus) others are of the opinion that 2 or more of the 4 existing wild species of fowl are responsible for the same.

Fowls are often classified based on the purpose for which they are developed such as egg type, meat type and dual purpose (for both egg and meat), but it is mostly on the basis of their origin. According to the latter, the birds are classified into the following major classes: American, Asiatic, English and Mediterranean. A breed refers to a group of domestic fowls with a common ancestry, and having similarity in shape, conformation, growth, temperament, shell colour of egg and breeds true to type. Variety is a subdivision of breed and within a breed there may be several varieties. The term variety is used to distinguish fowls having the characteristics of the breed to which they belong but differing in plumage colour, comb type, etc., from other groups of the same breed. A breed or a variety may have several strains or lines identified by a given name and produced by a breeder through at least 5 generations of closed flock breeding. Several strains within a breed or variety phenotypically may look alike but often differ in their production performance depending upon their breeding history.

- American class
- Asiatic class
- Mediterranean class
- English class

American Class

Dual-purpose Plymouth Rock, New Hampshire, Rhode Island Red and Wyandotte are the most popular American breeds. The American breeds are characterized by yellow skin, clean shanks free from feathers, red ear lobes and, except Lamona, lay brown-shelled eggs.

Plymouth Rock

Plymouth Rock is a much sought after American breed because of its egg size and fleshing properties. Barred and White Plymouth Rocks are very popular. Other varieties are Buff, Silver-pencilled, Partridge, Columbian and Blue. White Plymouth Rock with a long body of good depth and a broad and prominent breast is especially favoured for broiler production. The breed has a single comb. Standard weights (kg): Cock 4.3; hen 3.4; cockerel, 3.6; pullet, 2.7.

Rhode Island Red

Dual-purpose breed developed in Rhode Island in America, contains varying amounts of Malay games, Red Shanghasis, Brown Leghorn, Cornish and Wyandotte's blood. Single and Rose comb are the 2 common varieties. Some Single-comb Rhode Island Red strains are still very popular for commercial production of brown-shelled eggs. The most common colour is red buff; white and brown varieties are also found. The characteristic features of the breed are long body, broad and deep breast carried well forward, flat back with red eyes and red ear lobes. Legs and feet are deep yellow but may show brown colour. The male is dark red with black tail and shows black in both primary and secondary feathers of the wing when open. The female is even red with wing and tail marking as in male. Neck hackle shows little black marking at the base. Standard weights (kg): Cock, 3.8; hen, 2.9; cockerel, 3.4; pullet, 2.5.

Asiatic Class

Brahma, Cochin and Langshan, the 3 recognized Asiatic breeds which are virtually extinct now, are characterized by large body with heavy bones, feathered shanks, red ear lobes and yellow skin (except Black Langshan in which the skin is pinkish white). They are classed as broody and poor layers. These Asiatic breeds have contributed significantly to the development of American breeds.

Brahma

Brahma originated in the Brahmaputra Valley of India and is well known for its massive, well-feathered and proportioned body. Pea comb is one of the breed characteristics. Light, Dark and Buff are the most common varieties. Buff Brahma has plumage pattern similar to that of Columbian Plymouth Rock except that golden buff or buff is replaced by white. Its buff feathers are on shank and on the outer toe of each foot. Standard weights (kg) : Light Brahma-cock, 5.4; hen, .4.3; cockerel, 4.5; and pullet, 3.6. Dark Brahma-cock, 4.9; hen, 3.6; cockerel, 4.0; and pullet,3.1.

Cochin

Cochin also known as "Sanghai fowl" originated in Sanghai (China). It is characterized by massive appearance, thickly feathered shanks, single comb and cushion-like structure at the base of the tail. The popular varieties are Buff, Partridge, White and Black. Standard weights (kg): Cock, 4.9; hen, 3.8; cockerel, 3.61 pullet, 3.1.

Mediterranean Class

The Mediterranean breeds of Italian origin include Leghorn, Minorca, Andalusian, Spanish and Ancona. They are light bodied and are developed for high egg production. The Mediterranean breeds are characterized by white ear lobes, relatively large combs, non-broodiness, early maturity and white-shelled eggs.

Leghorn

Leghorn is characterized by compact and light body, uniform blending, pretty carriage, long shanks, small head with well set rose or single comb and early maturity. Popular varieties are White, Brown and black, of which White Leghorn is the most popular for its excellent laying performance. Standard weights (kg): White Leghorn: Cock, 2.6; hen, 2.0; cockerel, 2.2; pullet, 1.8.

English Class

The breeds/ of English origin are mostly utility breeds noted for their excellent fleshing properties. With the exception of Cornish, all the English breeds have white skin and red ear lobes. English breeds except Dorking and Red Cap lay brown-shelled eggs. All are classed as broody, but this defect can be gradually eliminated by selective breeding.

Cornish

Cornish, originally known as the Cornish Indian Game, appears to have been developed in England about the middle of the nineteenth century from crossings involving the Aseel, the Malay and English game breeds. It is noted for its close and compact feathering and heavy flesh with distinctive shape. Its breast is very deep and broad, giving the shoulders great width. All Cornish birds have pea comb. Standard weights (kg): Of the dark and white varieties are: Cock, 4.5; hen, 3.4; cockerel, 3.6; pullet, 2.7. The standard weights (kg) of the white-laced red variety are: Cock, 3.6; hen, 2.7; cockerel, 3.1; pullet, 22.

Australorp

This breed originated from the Black Orpington and was developed in Australia. It is more upstanding and less massive in appearance than the Black Orpington. It has been evolved as a layer bird. The back is rather long, with a gradual sweep to the tail. The body has good depth, but the feathering fits more closely to the body than it does in Orpington. The comb is single, beak is black, and shanks and toes are black or lead-black. The bottom of feet and toes are pinkish-white. The plumage is lustrous and greenish-black in all sections. The under-colour is dull black.

Standard weights (kg): Cock, 3.8; hen, 2.9; cockerel, 3.4; pullet, 2.5.

Indian Breeds

A large number of fowls of different sizes, shapes and colours, resembling the jungle fowls, are found all over India. Those with Chittagong, Aseel, Langshan or Brahma blood, are bigger in size and better in meat quality than the common fowls. Some Indian fowls resemble the Leghorn in size and shape, but are poor layers. One variety resembles the Sussex or Plymouth Rock in shape, but is smaller, lays fairly well and is more common in the eastern parts of the country. The common country hen, the desi is the best mother for hatching, a good forager but a poor layer. There are only 4 pure breeds of fowls indigenous to India. They are Aseel, Chittagong, Busra and Kadaknath.

Aseel

Aseel is noted for its pugnacity, high stamina, majestic gait and dogged fighting qualities. The best specimens of the breed, although rare, are encountered in parts of Andhra Pradesh, Uttar Pradesh and Rajasthan. The most popular varieties are Peela (golden red), Yakub (black and red), Nurie (white), Kagar (black), Chitta (black and white spotted), Java (black), Sabja (white and golden or black with yellow or silver), Teekar (brown) and Reza (light red). Although poor in productivity, they are known for their meat qualities. Broodiness is very common; the hen is a good sitter and efficient mother. They possess small and firmly set pea combs. Wattles and ear lobes are bright red. The beak is short. The face is long and slender, and not covered with feathers. The eyes are compact and well set. The neck is long, uniformly thick but not fleshy. The body is round and short with a broad breast, straight back and close-set strong tail root. The general feathering is close, scanty and almost absent on the breast. The plumage has practically no fluff and the feathers are tough. The tail is small and drooping. The legs are strong, straight, and set well apart. Standard weights (kg): Cock, 4 to 5; hen, 3 to 4; cockerel, 3.5 to 4.5; pullet, 2.5 to 3.5.

Kadaknath

The original name of the breed seems to be "Kalamasi", meaning a fowl with black flesh. These are bred by tribals in Jhabua and Dhar districts in western Madhya Pradesh. The eggs are light brown. The adult plumage varies from silver and gold-spangled to bluish-black without any spangling. The skin, beak, shanks, toes and soles of feet are slate-like in colour. The comb, wattles and tongue are purple.

Most of the internal organs also show intense black colouration which is pronounced in trachea, thoracic and abdominal air-sacs, gonads and at the base of the heart and mesentry. Varying degrees of black colouration are also seen in the skeletal muscles, tendons, nerves, meninges, brain, etc. The blood is darker than normal blood. The black pigment is due to the deposition of melanin.

This is a rather small-bodied bird with cocks weighing about 1.5

kg and hens 1.0 kg. The flesh, although repulsive to look at, is delicious. It lays about 80 eggs per year. It is resistant to diseases in its natural habitat in free range but is susceptible to Marek's disease under intensive rearing conditions.

Common Breeds of Duck

Indian Runner: This birds are found in some of the states in India, but they are bred pure in a very few places. The birds are white in colour. They carry their body in a noticeable upright position. The Indian Runner is widely used for crossing with the heavier breeds to improve their laying capacity. The egg is white in colour, average weight of an egg is over 56 grams.

Khaki Campbell

Khaki Campbell is the most popular breed, generally bred for profitable egg production in India. The colour of the drake is black at the neck and back. The duck is of Khaki colour which develops into grey when one year old. The ducks start laying eggs at about six months' age. The maximum numbers of eggs are laid during the second year. The birds are generally sold when four years old. The eggs are of white colour. The average weight of an egg is nearly 70 grams.

Chapter - 4

Economic Characters of Livestock and Poultry

There are certain inherited traits (characters) in various species of livestock which are economically important. These traits have to be improved for increased production like milk, meat, egg, wool, etc.

Economic traits of cattle and buffaloes:

1.	Age at first calving	:	Age in days of a cow of buffalo at the time of first calving
2.	Lactation length	:	Days in milk from the time of calving to the final drying off or cessation of milk yield.
3.	Lactation milk yield	:	Milk yield in kgs. from the date of calving to the date of drying.
4.	Dry period	:	Days from the date of first drying to the date of next calving
5.	305-day milk yield	:	Milk yield in kgs from the date of calving to the 305th day of lactation. If the lactation length is less than 305 days, the milk yield in kgs is considered as lactation yield for 305 days. When the lactation length is longer than 305 days, its milk yield in kgs is corrected to 305 days using correction factor.
6.	Calving interval or inter calving period	:	Days from the date of one calving to the date of next calving
7.	Peak yield	:	Highest daily yield recorded in kgs during lactation period.
8.	Average fat percentage	:	Average of fat tests done with milk samples drawn during a lactation at fortnight interval.
9.	Service period	:	It is the interval between date of calving

10.	Breeding efficiency	:	Measured as number of services per conception.
11.	Milk yield per day of lactation	:	Milk yield in kgs obtained by dividing total milk yield in a lactation by lactation length
12.	Milk yield per day of calving interval	:	It is milk yield in kgs obtained by dividing total milk yield in a lactation by calving interval.

Note: row 10 description begins with "and subsequent service resulting in fruitful conception." (continued from previous page)

List of registers to be maintained in cattle and buffalo breeding farms

Birth register	Semen evaluation register
Young stock register	Insemination register
Adult stock register	Milk recording register
Disposal register	Veterinary register
Mortality register	Feed register
Weighment register	Livestock account register
Service register for bull	Index cards for all calves/cow/bull
Semen collection register	

Economic Traits of Sheep

1.	Tupping percentage	:	Number of ewes mated to the number of ewes put to rams. This value is expressed in percentage.
2.	Lambing percentage	:	Number of ewes lambed to the number of ewes put to rams. This value is expressed in percentage.
3.	Weaning percentage	:	Number of lambs weaned to the number of ewes put to rams. This value is expressed in percentage.
4.	Twinning percentage	:	Number of twin births to the number of ewes lambed or to the total number of births (expressed in percentage).
5.	Birth weight	:	Weight in kg of a lamb at birth, weighed within 24 hours after birth.
6.	Weaning weight	:	Weight in kg of a lamb at 90th or 120th day of age.
7.	Weight at market age	:	Weight in kg at the age of 180, 240, 300 and 365th days. Usually for Indian sheep 9th month weight is called as market weight.
8.	Growth rate up to weaning	:	The rate of daily gain in weight from the date of birth to date of weaning.

Contd...

Economic Characters of Livestock and Poultry

9. Mortality percentage :

   Pre weaning mortality : Number of lambs died during the period zero to 90th day or up to 120th day to the number of lambs born alive.

   Post weaning mortality : Number of lambs died during the period of 90 or 120th day to 365 day to the total number of lambs available at weaning.

   Adult mortality : Number of adult sheep died to the number started at one year.

Wool Characteristics:

1. Grease fleece weight : Weight in kg of raw fleece shorn in a year or 6 months. Usually first at 6 months then annually shorn.

2. Clean fleece weight : Weight in kg of clean fleece derived from raw fleece shorn in a year or 6 months.

3. Fibre thickness or diameter or fineness of wool : The average diameter of wool in microns.

4. Fibre density : Average number of wool fibre in 1 sq. cm. Area.

5. Staple length : Length of wool in cm obtained by measuring the natural staple without stretching the crimps.

6. Medullation percentage : Number of medullated fibres to the total number of fibres examined.

List of registers to be maintained:

Birth register	Ewe performance register
Youngstock register	Wool register
Adult stock register	Mating register
Disposal register	Veterinary register
Mortality register	Feed register
Weighment register - Young stock and adult stock	Livestock account register
Ram performance register	

Economic Traits of Goat:

1. Age at first kidding : Age of the doe in days from the date of birth to the date of its first kidding.

2. Lactation length : Duration of lactation in days from the date

Contd...

3.	Lactation milk yield	:	of kidding to the date of drying off or cessation of lactation.
			Milk yield in kgs for the entire lactation.
4.	150 day lactation milk yield	:	Milk produced in kg up to 150 days of lactation. If the length is less than 150 days, it is taken as 150 days milk yield.
5.	Kidding interval	:	Interval in days between two successive kiddings.
6.	Incidence of multiple birth	:	Number of twin birth or triplet birth to the total number of does kidded.
7.	Growth performance traits	:	
	Birth weight	:	It is the weight in kg of a kid at birth, weighed within 24 hours after birth.
	Weaning weight	:	Weight in kg of a kid at 90 days of age.
	Weight at market age	:	Weight in kg. at the age of 6, 9 and 12th months of age.
8.	Mortality percentage	:	
	Pre-weaning mortality	:	No. of kids died during the period zero to 90th days to the number of kids born alive.
	Post weaning mortality	:	Number of lambs died during the period of 90 to 365 day to the total number of kids available at weaning.
	Adult mortality	:	Number of adult goats died to the number started at one year (in percentage).

List of registers to be maintained :

Birth register	Buck performance register
Young stock register	Mating register
Adult stock register	Milk recording register
Disposal register	Veterinary register
Mortality register	Feed register
Weighment register-Young stock and adult stock	Livestock account register

Economic Traits of Swine

Litter size at birth: Number of litters born in a litter.

Litter size born alive: Number of piglets born alive.

Litter size at weaning: Number of litters weaned per litter.

Weight at birth: Weight of a piglet in kg at birth.

Weaning weight: Weight of a piglet in kg at 56th day of age. If early weaning is practised, it is the weight at 28th day.

Weight at 154th day: Weight in kg at 154th day of age. (It is the first market age).

Weight at 210th day: Weight in kg at 210th day which is the second market age.

Growth Rate

1. Pre-weaning growth rate: Average daily gain in weight (in kg) from birth to weaning.

$$= \frac{Weaning\ weight\ (28/56^{th}\ day) - birth\ weight}{28/56}$$

2. Weaning to 154 day age: Average daily gain in weight (in kg) from weaning to 154 days.

$$= \frac{Weight\ 154^{th}\ day - Weaning\ weight\ (28/56^{th}\ day)}{182/152}$$

3. Weaning to 210 days: Average daily gain in weight (in kg) from weaning to 210th days.

$$= \frac{Weight\ at\ 210\ days - Weaning\ weight\ (28/56^{th}\ day)}{182/152}$$

Feed Efficiency

Weaning to 154 days: Weight in kg of feed consumed for producing one kg of live weight for the period from weaning to 154th day.

Weaning to 210 days: Weight in kg of feed consumed for producing one kg of live weight for the period from weaning to 210 days of age.

10 Mortality percentages (as percentage)

Pre-weaning mortality: No. of piglets died during the period birth to weaning (28/56 days) the number of piglets born alive.

Post weaning mortality: Number of piglets died during the period from 28/56th to 365 day to the total number of piglets available at weaning.

Adult mortality: Number of adult pigs died to the number started at one year (in percentage).

List of registers to be maintained:
- Birth register
- Youngstock register
- Adult stock register
- Disposal register
- Mortality register
- Weighment register
- Farrowing and growth record
- Service register
- Boar performance register
- Sow performance register
- Veterinary register
- Feed efficiency register
- Account register

CHAPTER - 5

Basic Statistics

Introduction

Statistics is a field of mathematics that pertains to data analysis. Statistical methods and equations can be applied to a data set in order to analyze and interpret results, explain variations in the data, or predict future data. A few examples of statistical information we can calculate are:

Average value (mean)

Most Frequently Occurring Value (Mode)

On average, how much each measurement deviates from the mean (standard deviation of the mean). Span of values over which your data set occurs (range), and midpoint between the lowest and highest value of the set (median)

Statistics is important in the field of engineering by it provides tools to analyze collected data. For example, a chemical engineer may wish to analyze temperature measurements from a mixing tank. Statistical methods can be used to determine how reliable and reproducible the temperature measurements are, how much the temperature varies within the data set, what future temperatures of the tank may be, and how confident the engineer can be in the temperature measurements made. This article will cover the basic statistical functions of mean, median, mode, standard deviation of the mean, weighted averages and standard deviations, correlation coefficients, z-scores, and p-values.

What is a Statistic?

In the mind of a statistician, the world consists of populations and samples. An example of a population is all 7th graders in the United States. A related example of a sample would be a group of 7th graders in the United States. In this particular example, a federal health care administrator would like to know the average weight of 7th graders and how that compares to other countries. Unfortunately, it is too expensive to measure the weight of every 7th grader in the United States. Instead statistical methodologies can be used to estimate the average weight of 7th graders in the United States by measure the weights of a sample (or multiple samples) of 7th graders.

Parameters are to Populations as Statistics are to Samples

A parameter is a property of a population. As illustrated in the example above, most of the time it is infeasible to directly measure a population parameter. Instead a sample must be taken and statistic for the sample is calculated. This statistic can be used to estimate the population parameter. (A branch of statistics knows as Inferential Statistics involves using samples to infer information about a population.) In the example about the population parameter is the average weight of all 7th graders in the United States and the sample statistic is the average weight of a group of 7th graders.

A large number of statistical inference techniques require samples to be a single random sample and independently gathers. In short, this allows statistics to be treated as random variables. A in-depth discussion of these consequences is beyond the scope of this text. It is also important to note that statistics can be flawed due to large variance, bias, inconsistency and other errors that may arise during sampling. Whenever performing over reviewing statistical analysis, a skeptical eye is always valuable.

Statistics take on many forms. Examples of statistics can be seen below:

When performing statistical analysis on a set of data, the mean, median, mode, and standard deviation are all helpful values to calculate. The mean, median and mode are all estimates of where the "middle" of a set of data is. These values are useful when creating groups or bins to organize larger sets of data. The standard deviation is the average distance between the actual data and the mean.

Mean and Weighted Average

The mean (also know as average), is obtained by dividing the sum of observed values by the number of observations, n. Although data points fall above, below, or on the mean, it can be considered a good estimate for predicting subsequent data points. The formula for the mean is given below as equation (1).

$$\overline{X} = \frac{\sum_{i=1}^{i=n} X_i}{n} \qquad (1)$$

However, equation (1) can only be used when the error associated with each measurement is the same or unknown. Otherwise, the weighted average, which incorporates the standard deviation, should be calculated using equation (2) below.

$$X_{wav} = \frac{\sum w_i x_i}{\sum w_i} \qquad (2)$$

where $w_i = \frac{1}{\sigma_i^2}$ and x_i is the data value.

Median

The median is the middle value of a set of data containing an odd number of values, or the average of the two middle values of a set of data with an even number of values. The median is especially helpful when separating data into two equal sized bins.

Mode

The mode of a set of data is the value which occurs most frequently.

Considerations

Now that we've discussed some different ways in which you can describe a data set, you might be wondering when to use each way. Well, if all the data points are relatively close together, the average gives you a good idea as to what the points are closest to. If on the other hand, almost all the points falls close to one, or a group of close values, but occasionally a value that differs greatly can be seen, then the mode might be more accurate for describing this system, whereas the mean would incorporate the occasional outlying data. The median is useful if you are interested in the range of values your system could be operating in. Half the values should be above and half the values should be below, so you have an idea of where the middle operating point is.

Standard Deviation and Weighted Standard Deviation

The standard deviation gives an idea of how close the entire set of data is to the average value. Data sets with a small standard deviation have tightly grouped, precise data. Data sets with large standard deviations have data spread out over a wide range of values. The formula for standard deviation is given below as equation (3).

$$\sigma = \sqrt{\frac{1}{n-1}\sum_{i=1}^{i=n}(X_i - \overline{X})^2} \qquad (3)$$

Side Note: Bias Estimate of Population Variance

The standard deviation (the square root of variance) of a sample can be used to estimate a population's true variance. Equation (3) above is an unbias estimate of population variance. Equation (3.1) below is another common method for calculating sample standard deviation, although it is an bias estimate. Although the estimate is biased, it is advantageous in certain situations because the estimate has a lower variance. (This relates to the bias-variance trade-off for estimators.) (3.1)

When calculated standard deviation values associated with weighted averages, equation (4) below should be used.

$$\sigma_{wav} = \frac{1}{\sqrt{\Sigma wi}} \qquad (4)$$

The Sampling Distribution and Standard Deviation of the Mean

Population parameters follow all types of distributions, some are normal, others are skewed like the F-distribution and some don't even have defined moments (mean, variance, etc.) like the Chaucy distribution. However, many statistical methodologies, like a z-test (discussed later in this article), are based off of the normal distribution. How does this work? Most sample data are not normally distributed.

This highlights a common misunderstanding of those new to

statistical inference. The distribution of the population parameter of interest and the sampling distribution are not the same. Sampling distribution?!? What is that?

Imagine an engineering is estimating the mean weight of widgets produced in a large batch. The engineer measures the weight of N widgets and calculates the mean. So far, one sample has been taken. The engineer then takes another sample, and another and another continues until a very larger number of samples and thus a larger number of mean sample weights (assume the batch of widgets being sampled from is near infinite for simplicity) have been gathered. The engineer has generated a sample distribution.

As the name suggested, a sample distribution is simply a distribution of a particular statistic (calculated for a sample with a set size) for a particular population. In this example, the statistic is mean widget weight and the sample size is N. If the engineer were to plot a histogram of the mean widget weights, he/she would see a bell-shaped distribution. This is because the Central Limit Theorem guarantees that as the sample size approaches infinity, the sampling distributions of statistics calculated from said samples approach the normal distribution.

Conveniently, there is a relationship between sample standard deviation (ó) and the standard deviation of the sampling distribution ($\sigma_{\bar{X}}$ - also know as the standard deviation of the mean or standard error deviation). This relationship is shown in equation (5) below:

$$\sigma_{\bar{X}} = \frac{\sigma_X}{\sqrt{N}} \qquad (5)$$

An important feature of the standard deviation of the mean, $\sigma_{\bar{X}}$ is the factor $\sqrt{N}$ in the denominator. As sample size increases, the standard deviation of the mean decrease while the standard deviation, ó does not change appreciably.

Example

You obtain the following data points and want to analyze them using basic statistical methods. {1,2,2,3,5}

Calculate the average: Count the number of data points to obtain n = 5

$$mean = \frac{1+2+2+3+5}{5} = 2.6$$

Obtain the mode: Either using the excel syntax of the previous tutorial, or by looking at the data set, one can notice that there are two 2's, and no multiples of other data points, meaning the 2 is the mode.

Obtain the median: Knowing the n=5, the halfway point should be the third (middle) number in a list of the data points listed in ascending or descending order. Seeing as how the numbers are already listed in ascending order, the third number is 2, so the median is 2.

Calculate the standard deviation: Using the equation shown above,

$$\sigma = \sqrt{\frac{1}{5-1}\left((1-2.6)^2 + (2-2.6)^2 + (2-2.6)^2 + (3-2.6)^2 + (5-2.6)^2\right)} = 1.52$$

Example *(Weighted)*

Three University of Michigan students measured the attendance in the same Process Controls class several times. Their three answers were (all in units people):

Student 1: A = 100 ± 3

Student 2: A = 105 ± 4

Student 3: A = 102 ± 2

What is the best estimate for the attendance A?

$$w_i = \frac{1}{\sigma_i^2} \quad w_1 = \frac{1}{9} \quad w_2 = \frac{1}{16} \quad w_3 = \frac{1}{4}$$

$$A_{wav} = \frac{\Sigma w_i A_i}{\Sigma w_i} = \frac{\frac{1}{9}*100 + \frac{1}{16}*105 + \frac{1}{4}*102}{\frac{1}{9} + \frac{1}{16} + \frac{1}{4}} = 101.92 \quad \text{students}$$

$$\sigma_{wav} = \frac{1}{\sqrt{\Sigma wi}} = \frac{1}{\sqrt{\frac{1}{9} + \frac{1}{16} + \frac{1}{4}}} = 0.65$$

Therefore,

A = 101.92 ± .65 students

Correlation Coefficient (r value)

The linear correlation coefficient is a test that can be used to see if there is a linear relationship between two variables. For example, it is useful if a linear equation is compared to experimental points. The following equation is used:

$$r = \frac{\Sigma(X_i - X_{mean})(Y_i - Y_{mean})}{\sqrt{\Sigma(X_i - X_{mean})^2}\sqrt{(Y_i - Y_{mean})^2}}$$

The range of r is from -1 to 1. If the r value is close to -1 then the relationship is considered anti-correlated, or has a negative slope. If the value is close to 1 then the relationship is considered correlated or to have a positive slope. As the r value deviates from either of these values and approaches zero, the points are considered to become less correlated and eventually are uncorrelated.

There are also probability tables that can be used to show the significant of linearity based on the number of measurements. If the probability is less than 5% the correlation is considered significant.

Linear Regression

The correlation coefficient is used to determine whether or not there is a correlation within your data set. Once a correlation has been established, the actual relationship can be determined by carrying out a linear regression. The first step in performing a linear regression is calculating the slope and intercept:

$$Slope = \frac{n\Sigma_i X_i Y_i - \Sigma_i X_i \Sigma_j Y_j}{n\Sigma_i X_i^2 - (\Sigma_i X_i)^2}$$

$$Intercept = \frac{(\Sigma_i X_i^2)\Sigma_i(Y_i) - \Sigma_i X_i \Sigma_i X_i Y_i}{n(\Sigma_i X_i^2) - (\Sigma_i X_i)^2}$$

Once the slope and intercept are calculated, the uncertainty within the linear regression needs to be applied. To calculate the uncertainty, the standard error for the regression line needs to be calculated.

$$S = \sqrt{\frac{1}{n-2}\left(\left(\sum_i Y_i^2\right) - intercept \sum Y_i - slope\left(\sum Y_i X_i\right)\right)}$$

The standard error can then be used to find the specific error associated with the slope and intercept:

$$S_{slope} = S\sqrt{\frac{n}{n\sum_i X_i^2 - (\sum_i X_i)^2}}$$

$$S_{intercept} = S\sqrt{\frac{\sum(X_i^2)}{n(\sum X_i^2) - (\sum_i X_i Y_i)^2}}$$

Once the error associated with the slope and intercept are determined a confidence interval needs to be applied to the error. A confidence interval indicates the likelihood of any given data point, in the set of data points, falling inside the boundaries of the uncertainty.

$$\beta = slope \pm \Delta slope \simeq slope \pm t^* S_{slope}$$

$$\alpha = intercept \pm \Delta intercept \simeq intercept \pm t^* S_{intercept}$$

Now that the slope, intercept, and their respective uncertainties have been calculated, the equation for the linear regression can be determined.

Y = âX + á

Z-Scores

A z-score (also known as z-value, standard score, or normal score) is a measure of the divergence of an individual experimental result from the most probable result, the mean. Z is expressed in terms of the number of standard deviations from the mean value.

$$z = \frac{X - \mu}{\sigma}$$

(6)

X = ExperimentalValue

μ = Mean

σ = StandardDeviation

Z-scores assuming the sampling distribution of the test statistic (mean in most cases) is normal and transform the sampling distribution into a standard normal distribution. As explained above in the section on sampling distributions, the standard deviation of a sampling distribution depends on the number of samples. Equation (6) is to be used to compare results to one another, whereas equation (7) is to be used when performing inference about the population.

Whenever using z-scores it is important to remember a few things:

Z-scores normalize the sampling distribution for meaningful comparison.

Z-scores require a large amount of data.

Z-scores require independent, random data.

$$\tilde{z}_{obs} = \frac{X - \mu}{\frac{\sigma}{\sqrt{n}}} \qquad (7)$$

n = SampleNumber

P-Value

A p-value is a statistical value that details how much evidence there is to reject the most common explanation for the data set. It can be considered to be the probability of obtaining a result at least as extreme as the one observed, given that the null hypothesis is true. In chemical engineering, the p-value is often used to analyze marginal conditions of a system, in which case the p-value is the probability that the null hypothesis is true.

The null hypothesis is considered to be the most plausible scenario that can explain a set of data. The most common null hypothesis is that the data is completely random, that there is no relationship between two system results. The null hypothesis is always assumed to be true unless proven otherwise. An alternative hypothesis predicts the opposite of the null hypothesis and is said to be true if the null hypothesis is proven to be false.

The following is an example of these two hypotheses:4 students who sat at the same table during in an exam all got perfect scores. Null Hypothesis: The lack of a score deviation happened by chance.Alternative Hypothesis: There is some other reason that they all received the same score.

If it is found that the null hypothesis is true then the Honor Council will not need to be involved. However, if the alternative hypothesis is found to be true then more studies will need to be done in order to prove this hypothesis and learn more about the situation.

As mentioned previously, the p-value can be used to analyze marginal conditions. In this case, the null hypothesis is that there is no relationship between the variables controlling the data set. For example:

Runny feed has no impact on product qualityPoints on a control chart are all drawn from the same distribution.Two shipments of feed are statistically the same.The p-value proves or disproves the null hypothesis based on its significance. A p-value is said to be significant if it is less than the level of significance, which is commonly 5%, 1% or .1%, depending on how accurate the data must be or stringent the standards are. For example, a health care company may have a lower level of significance because they have strict standards. If the p-value is considered significant (is less than the specified level of significance), the null hypothesis is false and more tests must be done to prove the alternative hypothesis.

Upon finding the p-value and subsequently coming to a conclusion to reject the Null Hypothesis or fail to reject the Null Hypothesis, there is also a possibility that the wrong decision can be made. If the decision is to reject the Null Hypothesis and in fact the Null Hypothesis is true, a type 1 error has occurred. The probability of a type one error is the same as the level of significance, so if the level of significance is 5%, "the probability of a type 1 error" is .05 or 5%. If the decision is to fail to reject the Null Hypothesis and in fact the Alternative Hypothesis is true, a type 2 error has just occurred. With respect to the type 2 error, if the Alternative Hypothesis is really true, another probability that is important to researchers is that of actually being able to detect this and reject the Null Hypothesis. This probability is known as the power (of the test) and it is defined as 1 - "probability of making a type 2 error."

If an error occurs in the previously mentioned example testing whether there is a relationship between the variables controlling the data set, either a type 1 or type 2 error could lead to a great deal of wasted product, or even a wildly out-of-control process. Therefore, when designing the parameters for hypothesis testing, researchers must heavily weigh their options for level of significance and power of the test. The sensitivity of the process, product, and standards for the product can all be sensitive to the smallest error.

Important Note About Significant P-values

If a P-value is greater than the applied level of significance, and the null hypothesis should not just be blindly accepted. Other tests should be performed in order to determine the true relationship between the variables which are being tested. More information on this and other misunderstandings related to P-values can be found at P-values.

Chi-Squared Test

A Chi-Squared test gives an estimate on the agreement between a set of observed data and a random set of data that you expected the measurements to fit. Since the observed values are continuous, the data must be broken down into bins that each contain some observed data. Bins can be chosen to have some sort of natural separation in the data. If none of these divisions exist, then the intervals can be chosen to be equally sized or some other criteria.

The calculated chi squared value can then be correlated to a probability using excel or published charts. Similar to the Fisher's exact, if this probability is greater than 0.05, the null hypothesis is true and the observed data is not significantly different than the random.

Calculating Chi Squared

The Chi squared calculation involves summing the distances between the observed and random data. Since this distance depends on the magnitude of the values, it is normalized by dividing by the random value

$$\chi^2 = \sum_{k=1}^{N} \frac{(observed - random)^2}{random}$$

or if the error on the observed value (sigma) is known or can be calculated:

$$\chi^2 = \sum_{k=1}^{N} \left(\frac{observed - theoretical}{sigma} \right)^2 x$$

Detailed Steps to Calculate Chi Squared.

Calculating Chi squared is very simple when defined in depth, and in step-by-step form can be readily utilized for the estimate on the agreement between a set of observed data and a random set of data that you expected the measurements to fit. Given the data:

	1	2	3	4	5	6	7
x_i	12	14	17	28	26	40	45
y_i	9	18	23	31	47	51	53
σ_i	3	4	2	4	4	3	5

Step 1: Find χ_o^2

$$\chi_o^2 = \sum_i \frac{(y_i - A - Bx_i)^2}{\sigma_{yi}^2}$$

When:

$$A = \frac{S_{xx}S_y - S_x S_{xy}}{\Delta}$$

$$B = \frac{SS_{xy} - S_x S_y}{\Delta}$$

$$\Delta = SS_{xx} - (S_x)^2$$

$$S = \sum \frac{1}{\sigma_{yi}^2}$$

$$S_x = \sum \frac{x_i}{\sigma_{yi}^2}$$

$$S_y = \sum \frac{y_i}{\sigma_{yi}^2}$$

$$S_{xy} = \sum \frac{x_i y_i}{\sigma_{yi}^2}$$

$$S_{xx} = \sum \frac{x_i^2}{\sigma_{yi}^2}$$

Step 2: Find the Degrees of Freedom

$$df = n - k$$

When: df = Degrees of Freedom
n = number of observations
k = the number of constraints

Step 3: Find $\tilde{\chi}_o^2 = \dfrac{\chi_o^2}{df}$

Basic Statistics

$\tilde{\chi}_o^2$ = the established value of $\tilde{\chi}^2$ obtained in an experiment with df degrees of freedom

Step 4: Find $P(\tilde{\chi}^2 \geq \tilde{\chi}_o^2)$ using published charts.

$P(\tilde{\chi}^2 \geq \tilde{\chi}_o^2)$ = the probability of getting a value of that is as large as the established

Step 5: Compare the probability to the significance level (i.e. 5% or 0.05), if this probability is greater than 0.05, the null hypothesis is true and the observed data is not significantly different than the random. A probability smaller than 0.05 is an indicator of independence and a significant difference from the random.

Chi Squared Test versus Fisher's Exact

For small sample sizes, the Chi Squared Test will not always produce an accurate probability. However, for a random null, the Fisher's exact, like its name, will always give an exact result.

Chi Squared will not be correct when:

fewer than 20 samples are being used

if an expected number is 5 or below and there are between 20 and 40 samples

For large contingency tables and expected distributions that are not random, the p-value from Fisher's Exact can be a difficult to compute, and Chi Squared Test will be easier to carry out.

Binning in Chi Squared and Fisher's Exact Tests

When performing various statistical analyzes you will find that Chi-squared and Fisher's exact tests may require binning, whereas ANOVA does not. Although there is no optimal choice for the number of bins (k), there are several formulas which can be used to calculate this number based on the sample size (N). One such example is listed below:

$k = 1 + \log_2 N$

Another method involves grouping the data into intervals of equal probability or equal width. The first approach in which the data is grouped into intervals of equal probability is generally more acceptable since it handles peaked data much better. As a stipulation, each bin should contain at least 5 or more data points, so certain adjacent bins sometimes need to be joined together for this condition to be satisfied.

Identifying the number the bins to use is important, but it is even more important to be able to note which situations call for binning. Some Chi-squared and Fisher's exact situations are listed below:

Analysis of a Continuous Variable

This situation will require binning. The idea is to divide the range of values of the variable into smaller intervals called bins.

Analysis of a Discrete Variable

Binning is unnecessary in this situation. For instance, a coin toss will result in two possible outcomes: heads or tails. In tossing ten coins, you can simply count the number of times you received each possible outcome. This approach is similar to choosing two bins, each containing one possible result. Examples of when to bin, and when not to bin.

Chapter - 6

Qualitative and Quantitative Inheritance

QUANTITATIVE VERSUS QUALITATIVE INHERITANCE

There is a continuum of traits being inherited as a Mendelian trait with simple inheritance and traits having quantitative inheritance without well separated classes and with many genes involved.

Classification of traits in relation to mode of inheritance and environmental tolerance are shown in Figure 1.2. First there are the well known traits with simple Mendelian mode of inheritance. The trait with quantitative genetic inheritance is caused by segregation of many gene pairs, each with small effect. At the same time the trait is influenced by a lot of minor environmental effects.

Diseases will often be 'either/or traits' as the simple Mendelian traits. Cases in which the severity of the disease has a normal distribution can also be found. In many production diseases the disease only occurs when a genetically prone individual is exposed to adverse environmental effects. Figure 1. Classification of traits in relation to mode of inheritance and environmental tolerance.

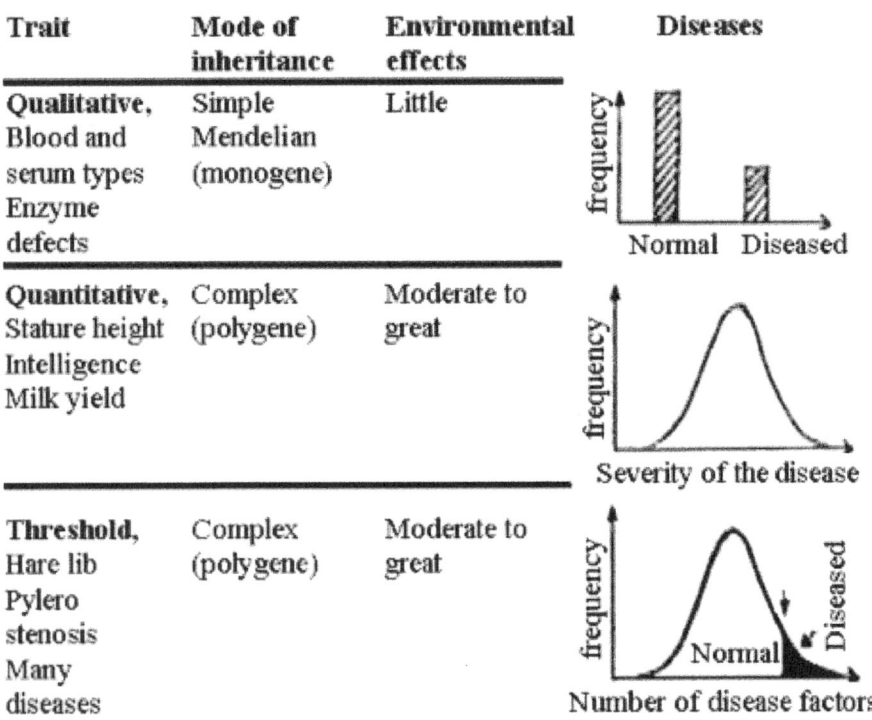

Traits are often grouped into those which exhibit qualitative differences and those which show quantitative differences.

Qualitative Traits

Qualitative traits are those which exhibit discontinuous variation i.e. those in which variations fall into a few clearly defined classes. The variations in quantitative traits, on the other hand, are continuous in that there are small gradations in expression from one extreme to the other.

Almost all of the traits whose inheritance is well known are in the qualitative class. The discontinuous nature of the variation in qualitative traits makes it possible, through breeding tests to determine the genotypes of individuals in each phenotypic class because-

- Most of the variation in qualitative traits can be attributed to one or a few major genes.
- These traits are little affected by environmental modifications. However, most traits of economic importance in farm livestock vary continuously. Major genes have not been found for such

quantitative traits as body weight, egg production, milk yield, or growth rate, much evidence exists to show that these traits are inherited. Some of this evidence is as follows:

- These traits responds to selection.
- Relatives resemble each other more closely than unrelated individuals for these traits.
- Breeds perform quite differently even when raised together and given the same care.

Difference between Qualitative and Quantitative Traits

	Qualitative traits	Quantitative traits
1.	Traits of kind with sharp distinction between phenotypes.	Traits of degree with no sharp distinction between phenotypes.
2.	Traits have discontinuous variation with discrete phenotypic classes.	Traits have continuous variation and phenotypic measurements from a spectrum.
3.	Traits are controlled by single pair of gene or only few gene pairs, effects of single genes on phenotype are noticeable (large). Such genes are called major genes or oligo genes.	Traits are controlled by many genes (perhaps 10 to 100 or more). Effects of single genes on phenotype are too small to be detected. Such genes are called polygenes or minor genes.
4.	Traits are usually not affected by environment.	Traits are affected by environment to a considerable extent.
5.	Inheritance of such traits is studied by making counts and ratios of phenotypes concerned with progeny of individual mating.	Complex statistical analysis are required to partition variance into components such as genetic and environmental, to study the inheritance of such traits.
6.	Few of such traits have economic importance.	Most of the economically important traits of farm animals are quantitative traits.
7.	Such traits show binominal distribution.	Traits show normal distribution.

Quantitative traits usually are characterized by Polygenic inheritance (i.e. genes at many loci) and large environmental effects. Although environmental effects can cause a discontinuous array of genotypes to appear continuous phenotypes, the inheritance of quantitative traits is usually polygenic. Major genes have not been detected for most quantitative traits despite intensive study. This supports the hypothesis that none of genes contributing to the overall genetic merit has sufficiently large effect to be detected, thus suggesting that a large number of loci are involved. The nature of weight, milk yield, etc. would suggest that many factors affecting the overall health, vigour, and metabolic process would also influence these traits.

Models for Quantitative Inheritance

All types of gene action probably are present among the genes affecting quantitative traits. Dominance and epistasis are almost certain to occur. Some evidence from inbreeding studies indicates that heterozygosity may be beneficial for some quantitative traits. The effects of genes are also greatly modified by the environment.

Additive model

The simplest model for quantitative inheritance is to assume that there are two alleles at each locus and the presence of each favourable allele(Capital letters) adds one unit more to genetic value than the alternative allele. The colour of the wheat Kernel provides as an example of this model as the genotypes below illustrate.

Genotypes	Phenotypes
AABB	Very dark red
AABb, AaBB	Dark red
AaBb, AAbb, aaBB	Medium red
Aabb, aaBb	Light red
aabb	White

Thus for each A or B genes which was substitute for an a or b gene the degree of redness increased one shade. If very dark red wheat were crossed with white the F_1 would all be medium red and the F_2 would exhibit the full range of variation in the following proportion: (1/16, 4/16, 6/16, 4/16, 1/16).

If we consider the same model with three contributing loci we get the following distribution of F_2 genotypes.

Genotypes	Phenotypes
6 favourable genes	1/64
5 favourable genes	6/64
4 favourable genes	15/64
3 favourable genes	20/64
2 favourable genes	15/64
1 favourable genes	6/64
0 favourable genes	1/64

Several things may be noticed from this model.

Average of the parent, F_1 and F_2 generations are identical. In fact, the average of the progeny from any mating is expected to be the same as the averages of the parents. As more loci are added the expected frequency of the extreme types in F_2 becomes smaller and smaller. Distribution of genetic values tends become more like a continuous distribution as more loci are added to the model. Genetic values tend toward a normal distribution as loci are added.

Dominance Model

A slightly more complex model is one where two dominant genes affect the trait in the following manner.

Genotypes	Size (arbitrary units)
Aabb, AAbb	3
AABB, AaBB, AABb, aabb	2
aaBb, aaBB	1

We see that presence of AA or Aa increases the size by one unit relative to aa and that BB or Bb decrease the size by one unit relatives to bb, i.e. A and B are dominant to a and b, respectively. If we make matings among these genotypes we get:

Parents	Expected progeny	Parental average	Progeny average
Aabb x Aabb	$\frac{1}{4}$ AAbb + $\frac{1}{2}$ Aabb + $\frac{1}{4}$ aabb	3	2.75
Aabb x aabb	$\frac{1}{2}$ Aabb + $\frac{1}{2}$ aabb	2.5	2.5
aaBb x AaBb	$\frac{1}{4}$ aaBB + $\frac{1}{2}$ aaBb + $\frac{1}{4}$ aabb	1.0	1.25
AaBb x AaBb	All nine genotypes	2.0	2.0

Many of these mating produce progeny which average the same as the average of the parents, but some do not. Upon closer examination we see that:

- Tall parents tend to produce progeny which average shorter than their parents.
- The progeny of short parents tend to average taller than their parents.

- Intermediate sized parents tend to produce progeny with more variation in size than parents for their extreme.
- If random mating occurs for several generations, the distribution of progeny genotypes will be the same as the distribution of parent genotypes.

❏❏❏

Chapter - 7

Gene and Genotype Frequency and Factors Influencing Gene Frequency

Gene Frequency

A population, in the genetic sense, is not just a group of individuals, but a breeding group and the genetics of a population is concerned not only with the genetic constitution of the individuals but also with the transmission of the genes from one generation to the next. In the transmission the genotypes of the parents are broken-down and a newest of genotypes is constituted in the progeny, from the genes transmitted in the gametes. The genes carried by the population thus have continuity from generation to generation, but the genotypes in which they appear do not. The genetic constitution of a population, referring to the genes it carries, is described by the array of gene frequencies, that is, by specification of the alleles present at every locus and the numbers of proportions of the different alleles at each locus. If for example, A_1 is an allele at the A locus, then the frequency of A_1, is the proportion or percentage of all genes at this locus that are the A_1 allele. The frequencies of all the allele at any one locus must add up to unity, or 100 percent.

The gene frequencies at a particular locus among a group of individuals can be determined from knowledge of the genotype frequencies. To take a hypothetical suppose there are two alleles A_1 and A_2, let the frequencies of genes and of genotypes be as follows:

	Genes		Genotypes		
	A_1	A_2	A_1A_1	A_1A_2	A_2A_2
Frequencies	P	q	P	H	R

So, that P + q and P+H+R = 1. Since each individual contains two genes, the frequency of A_1 genes is ½ (2 P+H) and the relationship between gene frequency and genotype frequency among the individuals counted is as follows:

$$P = P + \frac{1}{2} H$$

$$q = R + \frac{1}{2} H \ , \ P = \frac{2P + H}{2N} = \frac{P + \frac{1}{2} H}{N}$$

This proportion is known as the gene frequency of A in this group. Similarly, the frequency of the gene a in this group is

$$q = \frac{(H + 2R)}{2N} = \frac{\left(\frac{1}{2} H + R\right)}{N} \ , \text{ so that } p + q = 1.$$

Calculating gene frequency:

Autosomal loci with two alleles

(a) Co-dominant Autosomal Alleles

When co dominant alleles are present in a two allele system, each genotype has a distinctive phenotype. The numbers of each allele in both homozygous and heterozygous conditions may be counted in a sample of individuals from the population and expressed as a percentage of the total number of alleles in the sample. If the sample is representative of the entire population (continuing proportionally the same numbers of genotypes as found in the entire population) then we can obtain an estimate of the allele frequencies in the gene pool. Given a sample of N individuals of which D are homozygous for one allele (A^1A^1). H are heterozygous (A^1A^2) then N = D + H + R. since each of the N individuals are diploid at this locus, therefore

2N alleles represented in the sample letting P represent the frequency of the A^1 allele and the q the frequency of the A^2 allele, we have :

$$P = \frac{2D + H}{2N} = \frac{D + \frac{1}{2} H}{N} \ , \quad q = \frac{H + 2R}{2N} = \frac{\frac{1}{2} H + R}{N}$$

(b) Dominant and Recessive Autosomal Alleles

Determining the gene frequency for alleles which exhibit dominant and recessive relationship requires a different approach from that used with co dominant alleles. A dominant phenotype may have either of two genotypes, AA or Aa but we have no way of distinguishing how many are homozygous or heterozygous in our sample. The only phenotype whose genotype is known for certain is the recessive (aa). If the population is in equilibrium then we can obtain an estimate of q from q^2.

$$q = \sqrt{q^2} \qquad = P = 1 - q$$

(c) Sex influenced traits

The expression of dominance and recessive relationship may be changed in some genes when exposed to different environmental conditions. Most notable of which are the sex hormones. In sex influenced traits, the heterozygous genotype usually will produce different phenotypes in the two sexes, making the dominance and recessive relationships of the alleles appear to reverse themselves. We shall consider only those sex influenced traits whose controlling genes are on autosomes. Determination of allele frequencies must be indirectly made in one sex by taking the square root of the frequency of recessive phenotype $\left(q = \sqrt{q^2}\right)$. A similar approach for the opposite sex should give an estimate of P.

Autosomal Loci with Multiple Alleles

If we consider three alleles, A^1, A^2 and a with the dominance hierarchy A^1, $>A^2$, $>a$, occurring in the gene pool with respective frequencies P, q and r, then random mating will generate zygotes with the following frequencies:

$(P + q + r)^2 \qquad P^2 + 2Pq + 2Pr + q^2 + 2qr + r^2 = 1$

Genotype:	A1A1	A1A2	Aa	A2 A2	A2aA1a
Phenotype:		A1		A2	a

Sex Linked Genes

With sex linked genes the situation is rather more complex than with autosomal genes. The relationship between gene frequency and genotype frequency in the homogametic sex is the same as with an autosomal gene, but the heterogametic sex has only two genotype and each individual carries only one gene instead of two. For this reason two third of the sex linked genes in the population are carried by the homogametic sex and one third by the heterogametic sex will be referred to as male.

Consider two alleles A_1 and A_2 with frequencies P and Q and let the genotypic frequencies be as follows:

Frequencies	Female			Male	
	$A_1 A_1$	$A_1 A_2$	$A_2 A_2$	A_1	A_2
	P	H	Q	R	S

The frequency of A_1 among the female is than Pf = P + ½ H and the frequency among the males is Pm = R. The frequency of A_1 in the whole population is

$$\overline{P} = \frac{2}{3} P_F + \frac{1}{2} P_M$$

$$= \frac{1}{3}(2P_F + P_M)$$

$$= \frac{1}{3}(2P + H) + R$$

If the gene frequencies among males and among females are different, the population is not in equilibrium. The gene frequency in the population as whole does not change. Males get their sex linked genes only from their mother therefore p' m is equal to Pf in the previous generation. Female get their sex linked genes equally from both parents. Therefore p'f is equal to the mean of Pm and Pf in the previous generation using Primes to indicate the progeny generation. We have

$P'm = Pf$

$$P'f = \frac{1}{2}(Pm + Pf)$$

The difference between the frequencies in the two sexes is:

$$P'f - P'm = \frac{1}{2}(Pm + Pf) - Pf$$

$$= \frac{1}{2}Pm + \frac{1}{2}Pf - Pf$$

$$= \frac{1}{2}Pm - \frac{1}{2}Pf$$

$$= -\frac{1}{2}(Pf - Pm)$$

i.e. half the difference in the previous generation but in the other direction. Therefore the distribution of the genes between the two sexes oscillates, but the difference is halved in successive generations and the population rapidly approaches in equilibrium in which the frequencies in the two sexes are equal.

Autosomal loci with multiple alleles

If we consider three alleles, A, a' and a with the dominance hierarchy A a' a, occurring in the gene pool with respective frequencies P, q and r, then random mating will generate zygotes with the following frequencies:

$(P+q+r)^2 = P^2 + 2\ Pq + 2\ Pr + q^2 + 2\ qr + r^2 = 1$

Genotype:	AA	Aa Aa	a,a, Aa,	aa
Phenotype:		A	a,	a

Many multiple allelic series involve condominant relationship such as $(A_1 = A_2) > a$, with respective frequencies, P, q and r. More genotypes can be phenotypically recognized in codominant systems than in systems without codominance.

$(P+q+r)^2 = P^2 + 2\ Pr + 2\ Pq + q^2 + 2\ qr + r^2 = 1$

Genotypes:	$A_1 A_1$ $A_1 a$	$A_1 A_2$	$A_2 A_2$ $A_2 a$	aa
Phenotypes:	A_1	$A_1 A_2$	A_2	a

Random Mating

The random mating is that in the case of bisexual organism, any one individual of one sex is equally likely to mate with any individual of the opposite sex. In other words, the frequency of a certain type of mating is dictated by chance. Thus, if the mating is completely at random in the population (D, H, R), the frequency of AA x AA would be D^2 among all matings, assuming that genotypic proportions in both sexes are the same. The various types of matings and their frequencies are shown in table.

The term "panmixia" is sometimes used as a synonym of random mating and then the population is said to be panmictic.

Random mating frequencies

Females		Males		
		AA	Aa	aa
		D	H	R
AA	D	D^2	DH	DR
Aa	H	HD	H^2	HR
aa	R	RD	RH	R^2

Hardy – Weinberg Law

In a large random mating population with no selection mutation or migration, the gene frequencies and the genotype frequencies are constant from generation to generation, and further more, there is a simple relationship between the gene frequencies and the genotype frequencies. These properties of a population are derived from a theorem or principle, known as the Hardy -Weinberg law.

A population with constant gene and genotype frequencies is said to be in Hardy – Weinberg equilibrium. The relationship is this. If the gene frequencies of two alleles among the parents are P and q then the genotype frequencies among the progeny are P^2, 2 pq and q^2.

	Genes in parents		Genotypes in progeny		
Frequencies	A_1	A_2	A_1A_1	A_1A_2	A_2A_2
	P	Q	P^2	2pq	q^2

The conditions of random mating and no selection, required for the Hardy – Weinberg law to hold. Refer only to the genotypes under consideration. There may be preferential mating with respect to other attributes, and genotypes of other loci may be subject to selection, without affecting the issue. Two additional conditions are that the gene frequencies are the same in males and females. The reasons for these requirements will be seen in the proof. The proof of the Hardy – Weinberg law involves four steps.

1. Gene frequency in parents to gene frequency in gametes

Let the parent generation have gene and genotype frequencies as follows:

	Genes		Genotypes		
Frequencies	A_1	A_2	A_1A_1	A_1A_2	A_2A_2
	p	q	P	H	Q

Two types of genes produced, those bearing A_1 and A_2, A_1A_1 individuals produce only A_1 gametes A_1A_2 individuals provided segregation is normal, produce equal numbers of A_1 and A_2 gametes than provided all genotypes are equally fertile. The frequency of A_1 among all the gametes produced by the whole population is $P + ½ H$ is the gene frequency of A_1 in the parents producing the gametes.

2. From gene frequency in gametes to genotype frequencies in zygotes

Random mating between individuals in equivalent to random union among their gametes. The genotype frequencies among the zygotes (fertilized eggs) are than the products of the frequencies of the gametic types that unite to produce them. The genotype frequencies among the progeny produced by random mating can therefore the determined simple by the multiplying the frequencies of the gametic types produce by each sex of parents.

	Genotypes		
Frequencies	A_1A_1	A_1A_2	A_2A_2
	p^2	$2pq$	q^2

		Female gametes and their frequencies	
		$A_1(p)$	$A_2(q)$
Male gametes and their frequencies	A_1 (p)	A_1A_1 (p^2)	A_1A_2 (pq)
	A_2 (q)	A_1A_2 (pq)	A_2A_2 (q^2)

3. From zygotes to adults

The genotype frequencies in the zygotes deduced above are the Hardy – Weinberg frequencies. This, is, however, not quite the end of the proof because the frequencies will not be observable unless the zygotes survive equally well at least until they can be classified for genotype.

4. From genotype frequencies to gene frequency in progeny

This final step proves that the gene frequency has not changed. Provided the different genotypes in the progeny survive equally well to adulthood when they can become parents, their frequencies will be as above. The gene frequency in the adult progeny can be found by equation. The frequency of A_1 is $p^2+1/2\ (2Pq) = P^2+Pq = P(P+q) = P$ which is the same as in the parent generation.

Conditions

    Normal gene segregation
    Equal fertility of parents
    Equal fertilizing capacity of gametes
    Large population
    Random mating
    Equal gene frequencies in male and female parents
    Equal viability

Factors Influencing Gene Frequency

Hardy – Weinberg law with its assumption does not account for any change in gene frequency within populations. Soon after the Hardy – Weinberg's law was formulated it was felt that populations were not static as were assumed, but they changed due to changes in gene frequency. Mathematical explanations for such changes were necessary and they were provided by fisher, Wright and Haldane. Broadly there are two processes through which changes in gene frequency are brought about. They are systematic process and dispersive process (Falconer, 1967). In systematic process both the amount and direction of the changes can be predicted, while in the dispersive process the amount may be predictable but not the direction. The second type of changes occurs in all populations. In large and randomly mating populations dispersive process does not take place. The systematic process includes migration, mutation and selection as means to change gene frequency.

Migration

The rate of change in gene frequency in a population subject to immigration depends on the immigration rate and on the difference of gene frequency between immigrants and natives. Let us consider a large population consisting of a proportion of m of new immigrants in each generation, the remainar (1-M) being natives. Let us presume QM be the frequency of a certain gene among the immigrants and Qo among the natives. Then the frequency of the gene in the mixed population Q_1 will be as follows:

Q_1 = MQM + (1-M) Qo
   = MQM + Qo - MQo
   = M (QM - Qo) + Qo

The change in gene frequency after one generation of immigration will be the difference between the frequencies before and after immigration. The change of gene frequency is shown thus ΔQ

$\Delta Q = Q_1 - Qo$
= M (Qm - Qo) + Qo - Qo
= M (Qm - Qo)

Mutation

Mutations which change the gene frequencies are of two types – non-recurrent and recurrent. The non-recurrent mutations are rare and have little importance in population genetics as they do not being about perceptible changes in the gene frequencies. They have little chance to survive in a large population unless they have selective advantages. Thus a non-recurrent mutation without selective advantage cannot produce a permanent change in the population. It has equal chance for either to survive or to be lost. The loss is permanent and survival has little effect on change in gene frequency and thus it is taken as lost. The recurrent mutation changes the gene frequency. It recurs regularly with characteristic frequency and in the large population the frequency is never too low to be lost in sampling.

Let the gene A_1 mutates to A_2 gene per generation with a frequency of u per generation. Let the frequency of A_1 is po, so the new gene frequency is po - upo. If mutation occurs in both directions that is A_1 to A_2 and A_2 to A_1, then what will happen? For this, let us consider that po and qo be the initial gene frequency of A_1 and A_2 and that A_1 mutates to A_2 at a rate u and A_2 to A_1 at a rate v per generation than after one generation A_2 will gain by upo and loss by vqo. The change in gene frequency in one generation will be as shown below:

q = upo - vqo

this situation leads to a state of equilibrium where change in one allele is compensated by back mutation from the other. At this stage of equilibrium q = o and thus at equilibrium.

pu = qv

$$\frac{P}{q} = \frac{V}{u}$$

or $$\frac{1-q}{q} = \frac{V}{u}$$

or   u − uq = Vq

or   vq + uq = u

or   q (V+u) = u

or $$q = \frac{u}{V+u}$$

Selection

The individuals differ in viability and fertility. They, therefore contribute different numbers of offspring to the next generation. According to Falconer (1967), the proportionate contribution of off springs to the next generation is called the fitness of the individual, or sometimes the adaptive value or selective value. When the difference of fitness is due to presence or absence of a particular gene, the selection operatives on that gene. When selection operates on a particular gene, its frequency is changed in successive generation, because parents of different genotypes pass on their genes unequally to the next generation. Consequently gene and genotype frequencies are changed by selection. This change is brought about by reduced fertility, viability, culling or through eugenic measures. The intensity of selection is measured as co-efficient of selection (s) which is the proportionate reduction in the genetic contribution of a particular genotype in comparison to the standard genotype. The standard genotype is the most favoured genotype and its frequency is taken as one or unity. The contribution of the genotype selected against is 1 −s. For example, if selection coefficient against a particular genotype is 0.3, then out of 100 zygotes produced by favoured genotype, 70 per cent would be produced by the genotype against which selection operates. The fitness of a particular genotype is dependent upon the environment as well on genotypes at other loci. The fitness may vary from individual to individual. Hence average fitness of a particular gene is considered in a population. For arriving at a conclusion let us consider a pair of allelic gene A, and A_2 in which A_1 is dominant over A_2. Hence dominance means dominance with respect to fitness and not necessarily with respect to phenotypic effect of the gene. Let P and q be initial frequency of A_1 and A_2 respectively and s be the coefficient of selection. Wright has shown that the expected change in gene frequency "q for one generation of selection in which dominance is complete (ç=o) is

$$q = \frac{sq(1-q)^2}{1-S(1-q)^2}$$

For no dominance where ç = 1-2

$$q = \frac{sq(1-q)}{21-S(1-q)}$$

When the recessive gene is the desired one and ç = 1

$$q = \frac{sq^2(1-q)}{1-s(1-q^2)}$$

Genetic Drift

Genetic drift or allelic drift is the change in the frequency of a gene variant (allele) in a population due to random sampling. The alleles in the offspring are a sample of those in the parents, and chance has a role in determining whether a given individual survives and reproduces. A population's allele frequency is the fraction of the copies of one gene that share a particular form. Genetic drift may cause gene variants to disappear completely and thereby reduce genetic variation.

When there are few copies of an allele, the effect of genetic drift is larger, and when there are many copies the effect is smaller. Vigorous debates occurred over the relative importance of natural selection versus neutral processes, including genetic drift. Ronald Fisher held the view that genetic drift plays at the most a minor role in evolution, and this remained the dominant view for several decades. In 1968 Motoo Kimura rekindled the debate with his neutral theory of molecular evolution, which claims that most instances where a genetic change spreads across a population (although not necessarily changes in phenotypes) are caused by genetic drift.

Probability and Allele Frequency

The mechanisms of genetic drift can be illustrated with a simplified example. Consider a very large colony of bacteria isolated in a drop of solution. The bacteria are genetically identical except for a single gene with two alleles labeled A and B. Half the bacteria have allele A and the other half have allele B. Thus both A and B have allele frequency 1/2.

A and B are neutral alleles-meaning they do not affect the bacteria's ability to survive and reproduce. This being the case, all bacteria in this colony are equally likely to survive and reproduce. The drop of solution then shrinks until it has only enough food to sustain four bacteria. All the others die without reproducing. Among the four who survive, there are sixteen possible combinations for the A and B alleles:

(A-A-A-A),	(B-A-A-A),	(A-B-A-A),	(B-B-A-A),
(A-A-B-A),	(B-A-B-A),	(A-B-B-A),	(B-B-B-A),
(A-A-A-B),	(B-A-A-B),	(A-B-A-B),	(B-B-A-B),
(A-A-B-B)	(B-A-B-B),	(A-B-B-B),	(B-B-B-B).

If each of the combinations with the same number of A and B respectively are counted, we get the following table. The probabilities are calculated with the slightly faulty premise that the peak population size was infinite.

A		Combinations	Probability
4	0	1	1/16
3	1	4	4/16
2	2	6	6/16

The probability of any one possible combination is

$$\frac{1}{2} \cdot \frac{1}{2} \cdot \frac{1}{2} \cdot \frac{1}{2} = \frac{1}{16}$$

Where 1/2 (the probability of the A or B allele for each surviving bacterium) is multiplied four times (the total sample size, which in this example is the total number of surviving bacteria).

As seen in the table, the total number of possible combinations to have an equal (conserved) number of A and B alleles is six, and its probability is 6/16. The total number of possible alternative combinations is ten, and the probability of unequal number of A and B alleles is 10/16.

The total number of possible combinations can be represented as binomial coefficients and they can be derived from Pascal's triangle. The probability for any one of the possible combinations can be calculated with the formula

$$\binom{N}{k}(1/2)^N$$

where N is the number of bacteria and k is the number of A (or B) alleles in the combination. The function '()' signifies the binomial coefficient and can be expressed as "N chooses". Using the formula to calculate the probability that between them the surviving four bacteria have two A alleles and two B alleles.

$$\binom{4}{2}\left(\frac{1}{4}\right)^4 = 6 \cdot \frac{1}{16} = \frac{6}{16}$$

Genetic drift occurs when a population's allele frequencies change due to random events. In this example the population contracted to just four random survivors, a phenomenon known as population bottleneck. The original colony began with an equal distribution of A and B alleles but chances are that the remaining population of four members has an unequal distribution. The probability that this surviving population will undergo drift (10/16) is higher than the probability that it will remain the same (6/16).

Mathematical Models of Genetic Drift

Mathematical models of genetic drift can be designed using either branching processes or a diffusion equation describing changes in allele frequency in an idealised population.

Wright–Fisher Model

Consider a gene with two alleles, A or B. In diploid populations consisting of N individuals there are 2N copies of each gene. An individual can have two copies of the same allele or two different alleles. We can call the frequency of one allele p and the frequency of the other q. The Wright–Fisher model (named after Sewall Wright and Ronald Fisher) assumes that generations do not overlap. For example, annual plants have exactly one generation per year. Each copy of the gene found in the new generation is drawn independently at random from all copies of the gene in the old generation. The formula to calculate the probability of obtaining κ copies of an allele that had frequency p in the last generation is then $\frac{(2N)!}{k!(2N-k)!}p^k q^{2N-k}$

Where the symbol "!" signifies the factorial function. This expression can also be formulated using the binomial coefficient,

$$\binom{2N}{k} p^k q^{2N-k}$$

Moran Model

The Moran model assumes overlapping generations. At each time step, one individual is chosen to reproduce and one individual is chosen to die. So in each time step, the number of copies of a given allele can go up by one, go down by one, or can stay the same. This means that the transition matrix is tri diagonal, which means that mathematical solutions are easier for the Moran model than for the Wright-Fisher model. On the other hand, computer simulations are usually easier to perform using the Wright-Fisher model, because fewer time steps need to be calculated. In the Moran model, it takes N time steps to get through one generation, where N is the effective population size. In the Wright-Fisher model, it takes just one. In practice, the Moran model and Wright-Fisher model give qualitatively similar results, but genetic drift runs twice as fast in the Moran model.

Other Models of Drift

If the variance in the number of offspring is much greater than that given by the binomial distribution assumed by the Wright-Fisher model, then given the same overall speed of genetic drift (the variance effective population size), genetic drift is a less powerful force compared to selection. Even for the same variance, if higher moments of the offspring number distribution exceed those of the binomial distribution then again the force of genetic drift is substantially weakened.

Random Effects other than Sampling Error

Random changes in allele frequencies can also be caused by effects other than sampling error, for example random changes in selection pressure.

One important alternative source of stochasticity, perhaps more important than genetic drift is genetic draft. Genetic draft is the effect on a locus by selection on linked loci. The mathematical properties of genetic draft are different from those of genetic drift. The direction of the random change in allele frequency is auto correlated across generations.

Drift and Fixation

The Hardy-Weinberg principle states that within sufficiently large populations, the allele frequencies remain constant from one generation to the next unless the equilibrium is disturbed by migration, genetic mutation, or selection.

Populations do not gain new alleles from the random sampling of alleles passed to the next generation, but the sampling can cause an existing allele to disappear. Because random sampling can remove, but not replace, an allele, and because random declines or increases in allele frequency influence expected allele distributions for the next generation, genetic drift drives a population towards genetic uniformity over time. When an allele reaches a frequency of 1 (100%) it is said to be "fixed" in the population and when an allele reaches a frequency of 0 (0%) it is lost. Once an allele becomes fixed, genetic drift comes to a halt, and the allele frequency cannot change unless a new allele is introduced in the population via mutation or gene flow. Thus even while genetic drift is a random, directionless process, it acts to eliminate genetic variation over time.

Rate of allele frequency change due to drift

Assuming genetic drift is the only evolutionary force acting on an allele, after t generations in many replicated populations, starting with allele frequencies of p and q, the variance in allele frequency across those populations is

$$V_t \approx pq\left(1 - \exp\left\{-\frac{t}{2N_e}\right\}\right)$$

Time to Fixation or Loss

Assuming genetic drift is the only evolutionary force acting on an allele, at any given time the probability that an allele will eventually become fixed in the population is simply its frequency in the population at that time. For example, if the frequency p for allele A is 75% and the frequency q for allele B is 25%, then given unlimited time the probability A will ultimately become fixed in the population is 75% and the probability that B will become fixed is 25%. The expected number of generations for fixation to occur is proportional to the population size, such that fixation is predicted to occur much more rapidly in smaller populations. Normally the effective population size, which is smaller than the total population, is used to determine these probabilities. The effective population (N_e) takes into account factors such as the level of inbreeding, the stage of the lifecycle in which the population is the smallest, and the fact that some neutral genes are genetically linked to others that are under selection. The effective population size may not be the same for every gene in the same population. One forward-looking formula used for approximating the expected time before a neutral allele becomes fixed through genetic drift, according to the Wright–Fisher model, is

$$\bar{T}_{\text{fixed}} = \frac{-4N_e(1-p)\ln(1-p)}{p}$$

Where T is the number of generations, N_e is the effective population size, and p is the initial frequency for the given allele. The result is the number of generations <u>expected</u> to pass before fixation occurs for a given allele in a population with given size (N_e) and allele frequency (p). The expected time for the neutral allele to be lost through genetic drift can be calculated as

$$\bar{T}_{\text{lost}} = \frac{-4N_e p}{1-p} \ln p.$$

When a mutation appears only once in a population large enough for the initial frequency to be negligible, the formulas can be simplified to $\bar{T}_{\text{fixed}} = 4N_e$

for average number of generations expected before fixation of a neutral mutation, and $\bar{T}_{\text{lost}} = 2\left(\frac{N_e}{N}\right)\ln(2N)$ for the average number of generations expected before the loss of a neutral mutation.

Chapter - 8

Variation and Measurements

The amount of variation is measured and expressed as the variance. When values are expressed as deviations from the population mean the variance is simple the mean of the squared values. For example, the genotypic variance is the variance of genotypic values and the environmental variance is the variance of environmental deviations. The total variance is the phenotypic variance of the variance of phenotypic values, and is the sum of the separate components.

Components of Variance

The components of variance and the values whose variance they measure are listed as follows:

Variance component	Symbol	Value whose variance is measured
Phenotypic	V_P	Phenotypic value
Genotypic	V_G	Genotypic value
Additive	V_A	Breeding value
Dominance	V_D	Dominance deviation
Interaction	V_I	Interaction deviation
Environmental	V_E	Environmental deviation

The total variance is the phenotypic variance, or the variance of phenotypic values, and is the sum of the separate components. Thus

$$V_P = V_G + V_E$$
$$= V_A + V_D + V_I + V_E$$

But if genotypic values and environmental deviations are correlated, V_P will be increased by twice the covariance of G with E because if $P = G + E$, $V_P = V(G + E) = V_G + V_E + 2\,\text{Cov}_{GE}$. d. The term

2 Cov_{GE} is zero when G and E are uncorrelate also, when there is an interaction between genotypes and environments, there will be an additional component of variance due to the interaction i.e.

$V_P = V_G + V_E + V_I$ (G x E)

The partitioning of phenotypic variance (V_P) into genotypic (V_G) and environmental variance (V_E) is not sufficient for understanding the genetic properties of a population, particularly the cause of resemblance between relatives. Thus genotypic variance (V_G) must further be divided according to the division of genotypic value into breeding value, dominance deviation, and interaction deviation. Thus we have:

Value	G = A + D + I			
Variance Components	V_G (Genotypic)	= V_A (Additive)	+ V_D (Dominance)	+ V_I (Interaction)

The additive variance, which is the variance of breeding values, is obtained as follows:

The variance of the breeding values of three genotypes (AA, Aa, aa) is obtained by squaring the breeding values of three genotypes, multiplying by the frequency of the genotype concerned, and adding over the three genotypes. Thus,

$V_A = P^2 (2q\ a)^2 + 2pq\ [(q - p)\ a]^2 = q^2 (- 2p\ a)^2$
$= 4\ p^2q^2\ a^2 + 2pq\ (q - p)^2\ a^2 + 4\ p^2q^2\ á^2$
$= 2pq\ a^2\ (2pq + q^2 - 2pq + p^2 + 2pq)$
$= 2pq\ a^2\ (p^2 + 2pq + q^2)$
$= 2pq\ a^2$
$= 2pq\ (a + d\ (q - p))^2$

If there is no dominance at the locus under consideration (d = 0), than

$V_A = 2Pq\ a^2$

If there is complete dominance (d = a)

$V_A = 2pq\ (a + a\ (q - p))^2$
$= 2pq\ (a + a\ (q - 1 + q))^2$
$= 2pq\ (a + a\ (2q - 1))^2$
$= 2pq\ (a + 2qa - a)^2$
$= 2pq\ (2\ qa)^2$
$= 8\ pq^3a^2$

The variance of dominance deviation is

$$\begin{aligned} V_D &= P^2(-2q^2d)^2 + 2pq(2pqd)^2 + q^2(-2p^2d)^2 \\ &= 2p^2 q^4 d^2 + 8p^3 q^3 d^2 + 4p^4 q^2 d^2 \\ &= d^2(4q^4 q^2 + 8p^3 q^3 + 4p^4 q^2) \\ &= 4p^2 q^2 d^2 (q^2 + 2pq + p^2) \\ &= (2pqd)^2 \end{aligned}$$

If the gene frequencies are $p = q = 0.5$, the additive and dominance variances, with any degree of dominance becomes

$$V_A = 2 \times 0.5 \times 0.5 (a + d(0.5 - 0.5))^2$$

$$= \frac{1}{2} a^2$$

$$V_D = (2 \times 0.5 \times 0.5 \times d)^2$$

$$= \frac{1}{4} d^2$$

The total genotypic variance (V_G) is obtained as follows:

Since $G = A + D$

$$V_G = V_A + V_D + 2 \, Cov_{AD}$$

Where,

Cov_{AD} = The covariance of breeding values with deviation which is zero.

Hence

$$\begin{aligned} V_G &= V_A + V_D \\ &= 2pq(a + d(q - p))^2 + (2pqd)^2 \end{aligned}$$

Genes contribute much more variance when at intermediate frequencies than when at high or low frequencies in particular, recessive genes at low frequency, contribute very little variance. However, in practice we are not concerned with gene frequencies or gene effects while dealing variance components because gene frequencies or gene effects are not known except in specially constructed populations. Thus, we are concerned only with the estimation of the variance components. But it may be noted that as all the components of genetic variance are dependent on the gene frequencies, any estimate of the variance components are valid only for the populations from which they are estimated.

It may be mentioned here a possible misunderstanding which

may arise about the concept of additive genetic variance. The concept of additive variance does not carry with it the assumption of additive gene action i.e. the existence of additive variance is not an indication that any of the genes act additively (i.e. show neither dominance nor epistasis). Only if all the genotypic variance is additive can we conclude that the genes show neither dominance nor opistasis.

Interaction variance (V_I) is the variance of the interaction deviations, when the genotypes at different loci show epistatic interaction. Interactions involving large numbers of loci contribute so little variance that they can be ignored. Interaction variance is further subdivided according to whether the interaction involves breeding values or dominance deviations. Thus with two loci interactions there can be three kinds:

Interaction between two breeding values gives rise to additive x additive interaction variance, V_{AA}.

Interaction between the breeding values of one locus and the dominance deviation of the other and vice versa gives rise to additive x dominance interaction variance, V_{AD}.

Interaction between the two dominance deviations gives rise to dominance x dominance interaction variance, V_{DD}.

Thus,

$$V_I = V_{AA} + V_{AD} + V_{DD} + \text{etc.}$$

The term etc. refers to similar components arising from interactions involving more than two loci.

It is not easy to estimate the interaction variance. If it is not estimated separately, it is included with the dominance component of variance, which is then referred to as non – additive genetic variance.

Variance due to disequilibrium arises when a population is not in equilibrium under random mating. Disequilibrium exists when the genotype frequencies at two or more loci considered jointly are not what would be expected from the gene frequencies. the disequilibrium introduces an additional source of genetic variance due to the following reason.

Let there be two loci which do not interact in the manner discussed above. If G_1 and G_2 are the genotypic values of individuals with respect to each locus separately, and G, the genotypic value with respect to both loci jointly i.e.

$$G = G_1 + G_2$$

Then the total genotypic variance caused by the two loci together will be

$$V_G = V_{G_1} + V_{G_2} + 2Cov_{G_1 G_2}$$

The covariance term represents the correlation between the genotypic values at the two loci in different individuals. The correlation can be positive or negative. So disequilibrium can either increase or decrease the variance. When there is no disequilibrium, all the covariance terms are zero.

While partitioning phenotypic variance into genotypic variance and environmental deviations, it has been assumed that genotypic values and environmental deviations are independent of each other of there is no correlation between genotypic value and environmental deviation. Correlation is present when better genotypes are given better environments. e.g. normal practice of feeding cows according the their yield, introducing correlation between phenotypic value and environmental deviation, and since genotypic and phenotypic values are correlated, there is also a correlation between genotypic value and environmental deviation. When a correlation is present, the phenotypic variance is increased by twice the covariance of genotypic values and environmental deviation i.e.

$$V_P = V_G + V_E + 2\,Cov_{GE}$$

Neglecting the correlation between genotypic and environment leads to the covariance being included with the genotypic variance. Then this becomes a part of the individual's genotype.

Another assumption made while partitioning phenotypic variance into genotypic variance and environmental deviation, was that a specific difference of environment has the same effect on different genotypics. When this is not so, there is an interaction between genotypics and environments. e.g. change of environment 'X' to environment 'Y' have a greater effect on genotype 'A' than on 'B' or genotype 'A' may be superior to genotype 'B' in environment 'X', but inferior in environment 'Y'

When interaction between genotypes and environment is present, the phenotypic value of an individual is

$$P = G + E + I_{GE}$$

and

$$V_P = V_G + V_E + V_{GE}$$

Genotype – environment interaction becomes very important if individuals of a particular population are to be reared under different conditions.

Environmental variance (V_E) embraces all variation of non genetic origin, the nature of which depends very much on the character and the organism studied. Environmental variance is a source of error that reduces the precision in genetic studies and therefore the aim of the breeder is to reduce it as much as possible by careful management and proper design of experiments. Causes of environmental variations are:

- Nutritional factors
- Climatic factors
- Maternal effects
- Error of measurement
- Part of the 'intangiable' variation i.e. variation whose cause is unknown.

When more than one measurement of a character is possible on each individual e.g. milk yield, etc., the phenotypic variance can be portioned into variance within individuals and variance between individuals. Similarly environmental variance can also be portioned into environmental variance within individuals or special environmental variance, V_{ES}, arising from temporary or localized circumstances, and environmental variance between individuals or general environmental variance, V_{EG}, arising from permanent or non-localized circumstances. Use of this partitioning will be made in calculating repeatability to be dealt with later.

Chapter - 9

Values and Means of Population

The concept of value, expressible in the metric units by which the character is measured. The value observed when the character is measured on an individual is the phenotypic value of that individual. All observations of means, variance, or covariances, must clearly be based on measurements of phenotypic values. In order to analyse the genetic properties of the population we have to divide the phenotypic value into two componants.

The first division of phenotypic value is into components attributable to the influence of genotypic and environment. The genotype is the particular assemblage of genes possessed by the individual, and the environment is all the non-genetic circumstances under the term 'environment' means that the genotype and the environment are by definition the only determinants of phenotypic value. The two components of value associated with genotypic value and the environmental deviation.

$P = G + E$

Where,

P is the phenotypic value, G is the genotypic value and E is environmental deviation.

The mean environmental deviation in the population as a whole is taken to be zero. So that the mean phenotypic value is equal to the mean genotypic value. The term population mean than refers equally to phenotypic or to genotypic values. When dealing with successive generations we shall assume for simplicity that environment remains constant from generation to generation so that the population mean is constant in the absence of genetic change. If we replicate the a particular genotype in a number of individuals and measure them under environment conditions normal for the population, their mean environmental deviations would be zero and their mean phenotypic value would congiguently be equal to the genotypic value of that particular genotype.

In principle it is measurable, but in practice it is not, except when we are concerned with a single locus where the genotypes are phenotypically distinguishable or with the genotypes represented in highly inbred lines for example, considering a single locus with two alleles. A_1 and A_2, we call the genotypic value of one homozygote + a, that of the other homozygote - a and that of the heterozygote d (we shall adopt the convention that A_1 is the allele that increase the value. We thus have a scale of genotypic values as in fig. the origin or point of zero value on this scale is mid way between the values of the two homozygotes. The value d of the heterozygotes depends on the degree of dominance. If there is no dominance d = 0, if A_1 is dominant over A_2, d is positive and if A_2 is dominant over A_1, d is negative. If dominance is complete d is equal to + a or - a and if there is over dominance d is greater than + a or less than - a. The degree of dominance may be expressed as d/a.

Population mean

We can now see how the gene frequencies influence the mean of the character in the population. Let the gene frequencies of A_1 and A_2 be P and q respectively are as follows :

Genotype	Frequency	Value	Frequency x Value
$A_1 A_1$	p^2	+ a	$p^2 a$
$A_1 A_1$	2pq	d	2pqd
$A_2 A_2$	q^2	-a	$-q^2 a$
		Sum = a (p-q) + 2 dqp	

The mean value in the whole population is obtained by multiplying the value of each genotype by its frequency and summing over the there genotypes. The population mean, multiplication of the value by the frequency of each genotype is shown in the last column in the table. Summation of this column is simplifies by noting that $p^2 - q^1 = (p + q)$ the population mean, which is the sum of this column, is this M = a (p - q) + 2 dpq. This is both the mean genotypic value and the mean phenotypic value of the population with respect to the character.

The contribution of any locus to the population mean thus has two terms: a (p - q) attributable to homozygotes and 2 dpq attributable to heterozygotes. If there is nodominance

(d = 0) the second term is zero, and the mean is proportional to the gene frequency. M = a (1 - 2 q). If there is complete dominance (d = a), the mean is proportional to the square of the gene frequency = M = a (1 - 2 q²). The total range of values attributable to the locus is 2 a, in the absence of over dominance. That is to say, if A_1 were fixed in the population (p = 1). The population mean would be a, and if A_2 were fixed (q = 1) if would be - a. If the locus shows over dominance, however, the man of an unfixed population may be out side this range.

The genotypic values a and d are deviations from the mean value of the two homozygous. If the mean is to be expressed as deviation from some other value, an appropriate constant must be added or subtracted. For example, one might want to express the mean as a deviation from the value of the lower homozygotes. This would require the addition of a and the mean would become, after some simplification, M = 2P (a + dq) or expressed as deviation from the upper homozygote, it would be M = 2q (- a + dq) with additive combination, then the population mean resulting from the joint effects of several loci is the sum of the contributions of each of the separate loci thus:

M = Σa (p-q) + 2Σ d pq

This is again both the genotypic and phenotypic mean value. The total range in the absence of over dominance is now 2 Ó a. If all alleles that increase the value were fixed, the mean would be - Ó a. These are the theoretical limits to the range of potential variation in the population.

Average Effect

The new value associated with genes as distinct from genotypes is known as the average effect. Average effects depend on the genotypic values, a and d as previously defined and also on the gene frequencies. Average effects are therefore properties of populations as well as of the genes concerned. The concept of average effects is not easy to grasp. But it is fundamental to understanding the inheritance of quantitative characters. There are several ways in which average effects can be defined. One definition is this; the average effect of a particular gene (allels) is the mean deviation from the population mean of individuals which received that gene from one parent, the gene received from the other parent having come at random from the population. This may be stated in another way, let a number of gamets all carrying A_1 unite a random with gametes from the population, then the mean of the genotypes so produced deviates from the population mean by an amount which is the average effect of the A_1 gene are as follows.

Type of -gamete	Value and frequencies of genotype produced			Mean value of genotypes produced	Population mean to be deducted	Average effect of gene
	A_1A_1 d	A_1A_2 d	A_2A_2 -a			
A_1	P	q	-	Pa + qd	-[a(P-q)+2dpq]	q[a+d(q-P)]
A_2	-	P	q	-qa+Pd	-[a(P-q)+2dpq]	-P[a+d(q-P)]

Let us see how the average effect is related to the genotypic values a and d, in terms of which the population mean was expressed. Consider a locus with two alleles A_1 and A_2, at frequencies p and q respectively, and take first the average effect of the gene A_1, for which we will use the symbol. If gametes carrying A_1 unite at random with gametes from the population, the frequencies of the genotype produced will be P of $A_1 A_1$ and q of $A_1 A_2$. The genotypic value $A_1 A_1$ is and that of $A_1 A_2$ is d, and the mean of these, taking account of the proportions in which they occur, is Pa+qd. The difference between this mean value and population mean is the average effect of the gene A_1.

Taking the value of the population mean, we get

α_1 = Pa +qd - [a (P - q) + 2 dpq]

= q [a +d (q - P)]

Similarly, the average effect of the gene A_2 is

α_1= - P (a +d (q - P)

when only two alleles at a locus are under consideration it is more convenient to express their average effects in terms of the average effect of the gene substitution. This is simply the difference between the average effects of the two alleles, but its meaning may be more clearly understood in the following ways. Suppose that we could change A_2 genes chosen at random into A_1 genes, as if by direct mutation, and could then note the resulting change of value. The mean change so produced would be the average effect of the gene substitution. When A_2 genes are chosen at random a proportion P will be found in $A_1 A_2$ genotypes (P being the gene frequency of A_1) and a proportion q in $A_2 A_2$ genotypes. Changing $A_1 A_2$ into $A_1 A_1$ will change the value from d to +a and the effect will therefore be (a−d). The changing of $A_2 A_2$ into $A_1 A_2$ will change the value from −a to d, and the effect will be (d+a). The average change is therefore P (a−d) +q (d+a), which on rearrangement becomes a+d (q−P). Thus the average effect of the gene substitution is

$\alpha = a + d(q-P)$

it can readily be seen that

$\alpha = \alpha_1 - \alpha_2$

and that the average effects of the two alleles, when expressed in terms of the average effect of the gene substitution, are:

$\alpha_1 = q\alpha$
$\alpha_2 = P\alpha$

Breeding Value

The value of an individual, judged by the mean value of its progeny, is called the breeding value of the individual. Breeding value, unlike average effect, can therefore be measured. If an individual is mated to a number of individuals taken at random form the population, then its breeding value is twice the mean deviation of the progeny from the population mean. The deviation has to be doubled because the parent in question provides only half the genes in the progeny. The other half coming at random form the population. Breeding values can be expressed in the form of deviations from the population mean as defined above just as the average effect is a property of the gene and the population, so is the breeding value a property of the individual and the population from which its mates are drawn.

Defined in terms of average effects, the breeding value of an individual is equal to the sum of the average effects of the genes it carries, the summation being made over the pairs of alleles at each locus and overall loci, thus, for a single locus with two alleles, the breeding values of the genotypes are as follows:

Genotype	Breeding value
$A_1 A_1$	$2 á_1 = 2 q á$
$A_1 A_2$	$á_1 + á_2 = (q - P) á$
$A_2 A_3$	$2 á_2 = -2 P á$

The breeding value of any genotype being the sum of the average effects of the two alleles present. If all loci are to be taken into account, the breeding value of a particular genotype is the sum of the breeding values attributable to each of the separate loci.

Dominance Deviation

The difference between the genotypic value G and the breeding value of A of a particular genotype is known as the dominance deviation D, so that

$G = A + D$

The dominance deviations arises form the property of dominance among the alleles at a locus, since in the absence of dominance, breeding values and genotypic values coincide. Form the statistical point of view the dominance deviations are interaction between alleles, of within locus interactions. Since the average effects of genes and the breeding values of genotypes, depend on the gene frequency in the population. The dominance deviations are also dependent on gene frequency. They are therefore partly properties of the population and are not simply measures of the degree of dominance. The dominance deviation can be expressed in terms of the arbitrarily assigned genotypic value a and d, by subtraction of the breeding value from the genotypic value, as shown in table. The genotypic values must first be converted to deviations from the population mean. The genotypic values, converted, are given in two forms, in terms of a and in terms of á. The arbitrarily assigned genotypic value of A_1, A_2 is + a, and the population mean is (P − q) + 2 dPq. Expressed as a deviation from the population mean, the genotypic value is therefore

a − [a (P − q) + 2dPq] = a (− P + q) − 2 dPq = $2q^a$ − 2dPq = 2q (a − dP)

This may be expressed in terms of the average effect á by substituting a = á − d (q − P) and the genotypic value then becomes 2q (á − qd). Subtraction of the breeding value, 2qd, gives the breeding value, 2q á, gives the dominance deviation as − $2q^2d$. By similar reasoning the dominance deviation of $A_2 A_2$ is 2Pqd and that of $A_2 A_2$ is − $1p^2d$. Thus all the dominance deviations are also all zero. Therefore in the absence of dominance breeding values and genotypic values are the same. Genes that show no dominance (d=0) are sometimes called, additive genes, or are said to act additively.

We have given two definitions of breeding value, a practical one in terms of the measured value of the progeny and a theoretical one in terms of average effects. Non additive combination renders these two definitions not guite equivalent. In a population in Hardy-Weinberg equilibrium the mean breeding value must be equal to the mean genotype value and to the mean phenotypic value. Multiplying the breeding value by the frequency of each genotype and summing gives the mean breeding value (expressed as a deviation from the population mean) as:

$2P^2q\ \alpha + 2Pq\ (q - P)\ \alpha - 2q^2P\ \alpha = 2Pqd\ (P + q - P - q) = 0$

The breeding value is some times referred to as the additive genotype and variation in breeding value ascribed to the "additive effects" of genes.

Because the breeding value express the breeding value transmitted from parents to offspring to follows that the expected breeding value of any individual is the average of the breeding values of its two parents, and it follows from the definition of breeding value that this is also the individuals expected phenotypic value. The "expected" values are simply the mean values of a large number of offspring of the same parents. So that transmission of values from parents to offspring is expressed by

$$\overline{P}_o = \overline{A}_o = \frac{1}{2}(A_s + A_d)$$

Where, the subscripts o, s and d refer to offspring, sire and dam respectively.

Since the mean breeding value and the mean genotypic value are equal, it follows that the mean dominance deviation is zero. This can be verified by multiplying the dominance deviation by the frequency of each genotype and summing the mean dominance deviation is thus

$-2\ P^2q^2d + 4P^2q^2d - 2P^2q^2d = 0$

Interaction Deviation

When only a single locus is under consideration, the genetic value is made up of the breeding value and the dominance deviation only. But when the genotype refers to more than one locus, the genotypic value may contain an additional deviation due to non additive combination. Let G_A and G_B be the genotypic values of an individual attributable to one and second locus respectively and G the aggregate genotypic value to the both loci together. Then

$G = G_A + G_B + I_{AB}$

Where I_{AB} is the deviation from additive combination of these genotypic values. In dealing with population mean, we assume that I was zero for all combinations of genotypes. If I is not zero for any combinations of genes at different loci, those genes are said to "interact" or to exhibit "epistasis". The deviation I is called the interaction deviation or epistatatic deviation. If the interaction deviation is zero the genes concerned are said to "act additively" between loci. Thus "additive" action may mean two different things. Referred to genes at one locus it means the absence of dominance, and referred to genes at different loci it means the absence of epistasis. So far all loci together we can write

G = A + D + I

Where A is the sum of the breeding values attributable to the separate loci and D is the sum of the dominance deviations.

Chapter - 10

Heritability – Methods of Estimation and Use

Estimates of heritability use statistical analyses to help to identify the causes of differences between individuals. Because heritability is concerned with variance, it is necessarily an account of the differences between individuals in a population. Heritability can be univariate- examining a single trait – or multivariate – examining the genetic and environmental associations between multiple traits at once. This allows a test of the genetic overlap between different phenotypes: for instance hair colour and eye colour. Environment and genetics may also interact, and heritability analyses can test for and examine these interactions (GxE models).

A prerequisite for heritability analyses is that there is some population variation to account for. In practice, all traits vary and almost all traits show some heritability. For example, in a population with no diversity in hair colour, "heritability" of hair colour would be undefined. In populations with varying values of a trait, variance could be due to environment (hair dye for instance) or genetic differences, and heritability could vary from 0-100%.

This last point highlights the fact that heritability cannot take into account the effect of factors which are invariant in the population. Factors may be invariant if they are absent and don't exist in the population (e.g. no one has access to a particular antibiotic), or because they are omni-present (e.g. if everyone is drinking coffee).

The concept of heritability plays a central role in the psychology of individual differences. Heritability has two definitions. The first is a statistical definition, and it defines heritability as the proportion of phenotypic variance attributable to genetic variance. The second definition is more common "sensical". It defines heritability as the extent to which genetic individual differences contribute to individual differences in observed behavior (or phenotypic individual differences). You should memorize both of these definitions.

Because heritability is a proportion, its numerical value will range from 0.0 (genes do not contribute at all to phenotypic individual differences) to 1.0 (genes are the only reason for individual differences). For human behavior, almost all estimates of heritability are in the moderate range of .30 to .60.

The quantity (1.0 - heritability) gives the environmentability of the trait. Environmentability has an analogous interpretation to heritability. It is the proportion of phenotypic variance attributable to environmental variance or the extent to which individual differences in the environment contribute to individual differences in behavior. If the heritability of most human behaviors is in the range of .30 to .60, then the environmentability of most human behaviors will be in the range of .40 to .70.

The expression of a trait is a function of the hereditary potential of the organism and the environment in which it develops. This relationship may be illustrated as follows:

$$P = G + E + GE \quad (1)$$

Where P is the organism's phenotype, G is the genetic contribution, E is the environment and HE is any interaction between heredity and environment. Both variables, G and E, must remain within certain limits if the organism is to survive: thus, arguments of heredity vs. environment or nature vs. nurture regarding the expression of a trait are obviously pointless. However, the relative contribution of hereditary differences and environmental differences to phenotypic differences in a population is of practical interest. The magnitude of these phenotypic differences is measured by the phenotypic variance (σ_P^2) of the population. From equation (1), it is seen that phenotypic variance will include the hereditary variance (σ_G^2), the environmental variance (σ_E^2) and the variance of the interaction between heredity and environment (σ_{GE}^2) s follows:

$$\sigma_P^2 = \sigma_G^2 + \sigma_E^2 + \sigma_{GE}^2 \quad (2)$$

The fraction of the phenotypic variability of a trait that is due to

Heritability – Methods of Estimation and Use

hereditary differences is the "Heritability" of the trait. This measures the relative contribution of hereditary differences to phenotypic differences. Since only the fraction of the phenotypic superiority of parents that is due to heredity is transmitted to offspring, the relative magnitude of this fraction is of great practical significance in a breeding program.

It was shown previously that he hereditary variance (σ_G^2) may be subdivided as follows:

$$\sigma_G^2 = \sigma_A^2 + \sigma_D^2 + \sigma_I^2 \tag{3}$$

Where, σ_G^2 is the additive genetic variance, σ_D^2 is the dominance variance and σ_I^2 is the epistatic variance. Therefore, the fraction of the phenotypic variance due to all types of genetic variance should be as follows:

$$h^2{}_{(B)} = \frac{6_G^2}{6_P^2} = \frac{6_A^2 + 6_D^2 + 6_I^2}{6_A^2 + 6_D^2 + 6_I^2 + 6_E^2 + 6_{GE}^2}{}^2 \tag{4}$$

This ratio is called "heritability in the broad sense".

It has been previously stated that mass selection is most effective when the phenotypic variance is due primarily to additive genetic differences. The fraction of the phenotypic variance caused by those additive genetic differences is termed "heritability in the narrow sense" and is as fallow:

$$h^2{}_{(N)} = \frac{\sigma_A^2}{\sigma_P^2} \tag{5}$$

Because of the relevance of $h^2{}_{(N)}$ to the effectiveness of mass selection, it is usually attempted to estimate $h^2{}_{(N)}$; however, most methods for estimating h^2 yield values between $h^2{}_{(N)}$ and $h^2{}_{(B)}$. it was indicated that $h^2{}_{(N)}$ is really the regression of the additive genetic values (or breeding value) on the phenotypes.

Since h^2 is a function of both σ_G^2 and σ_E^2 it will change if σ_G^2 or σ_E^2 change. Thus, h^2 is a population concept which is relative to a particular population at a particular time.

Estimation of Heritability

As the heritability (in the narrow sense) is the ratio of the additive genetic variance to the total variance, the methods of its estimation is derived form measures of the resemblance between related animals because the additive genetic variance is the chief determinant of resemblance between relatives. Thus all the different methods of estimation of heritability are derived form estimates of either a regression coefficient or an intraclass correlation among related individuals.

Regression estimates of heritability:

The resemblance between offspring and parent is expressed as the regression of offspring on parent. The three regression estimates are:

Regression of offspring on dam (b_{OD})

Regression of offspring on sire (b_{OS})

Regression of offspring on mid-parent $\left(b_{O\overline{P}}\right)$

These regression estimates are expressed in terms of heritability as follows:

Relatives	Covariance	Regression (b)/ Heritability (h2)
Offspring and dam	1/2 VA	$b = \dfrac{1}{2}\dfrac{V_A}{V_P} = \dfrac{1}{2}h^2$
Offspring and sire	1/2 VA	$b = \dfrac{1}{2}\dfrac{V_A}{V_P} = \dfrac{1}{2}h^2$
Offspring and mid-parent	1/2 VA	$b = \dfrac{1}{2}\dfrac{V_A}{V_P} = \dfrac{1}{2}h^2$

If maternal effects are assumed absent, the expected genetic composition of all the three estimates is the same. In case of regression of offspring on parent (either dam or sire), heritability can be estimated by doubling the regression. Let the phenotypes of the parent and offspring be

$$X_1 = g_1 + e_1 \tag{6}$$

and

$$X_2 = g_2 + e_2 \tag{7}$$

The heritability is then given by 2 b_{OP} (either b_{OD} or b_{OS}).

$$h^2 = 2\frac{Cov(Parent, offspring)}{Var(Parent)}$$

$$= 2\frac{Cov(g_1 + e_1, g_2 + e_2)}{\sigma_P^2} \tag{8}$$

Assuming that the environmental parts of the two phenotypes (e_1 and e_2) have zero covariances and so also the genotype environment covariance, we have

$$h^2 = 2\frac{Cov(g_1, g_2)}{\sigma_P^2}$$

$$= 2\frac{\frac{1}{2}\sigma_A^2 + \frac{1}{4}\sigma_{AA}^2 + \frac{1}{8}\sigma_{AAA}^2 + \ldots}{\sigma_P^2}$$

$$= \frac{\sigma_A^2}{\sigma_P^2} + \frac{\frac{1}{2}\sigma_{AA}^2 + \frac{1}{4}\sigma_{AAA}^2 + \ldots}{\sigma_P^2} \tag{9}$$

Similarly, in case of regression of offspring on mid-parent $(b_{O\bar{P}})$, the genetic composition of the heritability estimate is as above. Let the two parental and offspring phenotypes be

$X_1 = g_1 + e_1$ ⎫
$X_2 = g_2 + e_2$ ⎬ Parents
$X_3 = g_3 + e_3$ Offspring

The heritability estimate is the regression $_{(...)}$ itself.

$$h^2 = \frac{Cov(mean\ of\ parent,\ offspring)}{Var(mean\ of\ parent)}$$

$$\frac{Cov\left(\frac{1}{2}(g_1 + e_1 + g_2 + e_2), g_3 + g_3\right)}{Var\left(\frac{1}{2}(P_1 + P_2)\right)} \tag{10}$$

Assuming here that the environmental covariance between

parents, parents and offspring and genotype – environment covariance as zero, we have

Assuming equal variance of P_1 and P_2 we have

$$h^2 = \frac{\frac{1}{2}(Cov(g_1, g_3) + Cov(g_2, g_e))}{\frac{1}{4}(\sigma_P^2 + \sigma_P^2)}$$

$$= \frac{\frac{1}{2}\left(\frac{1}{2}\sigma_A^2 + \frac{1}{4}\sigma_{AA}^2 + \frac{1}{8}\sigma_{AAA}^2 + \ldots\ldots\right) + \left(\frac{1}{2}\sigma_A^2 + \frac{1}{4}\sigma_{AA}^2 + \frac{1}{8}\sigma_{AAA}^2 + \ldots\ldots\right)}{\frac{1}{2}\sigma_P^2}$$

$$= \frac{\left(\frac{1}{2}\sigma_A^2 + \frac{1}{4}\sigma_{AA}^2 + \frac{1}{8}\sigma_{AAA}^2 + \ldots\ldots\right) + \left(\frac{1}{2}\sigma_A^2 + \frac{1}{4}\sigma_{AA}^2 + \frac{1}{8}\sigma_{AAA}^2 + \ldots\ldots\right)}{\sigma_P^2} \quad (11)$$

$$= \frac{\sigma_A^2}{\sigma_P^2} + \frac{\frac{1}{2}\sigma_{AA}^2 + \frac{1}{4}\sigma_{AAA}^2 + \ldots\ldots}{\sigma_P^2}$$

Thus the theoretical value of all the heritability estimates is equal to the paremeter plus a bias due to epistatic components of variance. Thus another assumption in this method of estimation is the negligible bias form epistasis. A third assumption is that the models used are appropriate. While assumption on epistasis is usually reasonable, the assumption of zero covariance between environments and genotype and environment may be invalidated by similar treatment of daughters and dam. For the assumption on model it is usual to remove classifiable sources of variation such as age, herd, season and sex effects, before using production records for heritability estimates. The idea is that since animals will be evaluated for selection on the basis of such adjusted records, the heritability estimate should also be based on similar data. The adjustment is usually carried out by using correction factors or by computing the regression within classes, such as herds. Its purpose is to convert the phenotypic records such that each can be described by the genotype of the individual concerned and a random unclassifiable portion.

When maternal effects can not be ignored, the b_{OS} estimate is more nearly unbiased as it remains unaffected by maternal effects. However, this estimate is less commonly used because fewer degrees of freedom are available and also b_{OS} estimate is not available for sex limited traits (traits limited to one sex). It may be noted that regression estimates are not biased by selection of parents. This is because when as a result of selection the variance among the parents is reduced, the covariance is also reduced to the same extent so that the regression of offspring on parent is unaltered. However, the precision of the estimates is lowered as the sampling variance of the estimate increases. The regression estimates are also least affected by the system of mating. In the regression analysis when there are several offsprings per parent, the problem could be tackled in two ways:

- The parents records are repeated with each progeny record. This can be done where there is no environmental correlation between progeny of one parent. Usually this is more suitable with animal data.

- Average out all the offspring for each parent for obtaining the regression. This method is valid when there is a perfect correlation (unity) among members of each progeny group. Becker (1967) gives examples of both the methods. Kempthorne and Tandon (1953) describe a method striking a balance between these two procedures by weighting each progeny group mean by a factor which depends on the number of progeny and on their correlation with each other.

Correlation Estimates of Heritability

1. Half-sib covariance (Cov_{HS})

The phenotype of paternal half-sibs whose dams are random members of the population would be

$$X_{ij} = \mu + S_i + e_{ij} \qquad (12)$$

Where, μ = population mean

S_i = the effect of the sire

e_{ij} = random unclassifiable remainder including random effects of dams.

A second progeny of the same sire but different dam (1) would be

$X_{ij} = \mu + S_i + e_{i1}$

Then the covariance between these half-sibs will be

Cov (X_{ij}, X_{i1}) = Cov $((\mu + S_i + e_{ij}), (\mu + S_i + e_{i1}))$
= Cov (S_i, S_i)

Thus, the variance component for sires from the above model estimates the covariance between half-sibs subject to fulfilling the following 3 assumptions:

- There is no covariance between the sire and random effects, such as might arise with assortative mating, or the separate rearing of different sire groups.
- There is no environmental covariance between members of a sire group.
- The sires are a random sample from the population in question, and they are enough in number of the meaningful estimation of a variance component.

The covariance between half-sibs (Cov$_{HS}$) can be expressed as

$$Cov\ (HS) = \sigma_S^2 = \frac{1}{4}\sigma_A^2 + \frac{1}{16}\sigma_{AA}^2 + \frac{1}{64}\sigma_{AAA}^2 \cdots \cdots \quad (13)$$

Thus the stops involved in the estimation of heritability by the half-sib method are:

- Obtain an estimate of sire variance (σ_S^2)

- Multiply (σ_S^2) by 4 to obtain an estimate of the additive variance plus some epistatic variance.

- Obtain an estimate of phenotypic variance (σ_P^2) as the sum of the sire and random error variances $(\sigma_S^2 + \sigma_e^2)$.

The estimation of these variances are obtained by analysis of variance as follows:

Source	d.f.	Mean square	Expected mean square
Among sires	S - 1	MSS	$(\sigma_S^2 + \sigma_e^2)$
Among progeny within sires	S (t - 1)	MSE	$(\sigma_S^2 + \sigma_e^2)$

The variances are estimated as:

$$\sigma_e^2 = MSE$$

$$\sigma_S^2 = (MSS - MSE)/t$$

The h² is estimated as;

$$h^2 = \frac{4\sigma_S^2}{\sigma_S^2 + \sigma_e^2}$$

$$= \frac{4\left(\frac{1}{4}\sigma_A^2 + \frac{1}{16}\sigma_{AA}^2 + \frac{1}{64}\sigma_{AAA}^2 + \ldots\ldots\right)}{\sigma_P^2} \quad (14)$$

Thus, the upward bias of the heritability estimate by half-sib method is less than in the parent offspring regression estimate. But in field data if there is some environmental variance confounded with the sire variance, this will be increased four limes by multiplication with 4 required in this method.

It may be noted that the ratio $\sigma_S^2/(\sigma_S^2 + \sigma_e^2)$ is the intraclass correlation described earlier. Heritability is obtained by multiplying this ratio with four.

Full-sib covariance (Cov$_{FS}$):

The models which would describe the occurrence of full-sib families may be:

Data using human or laboratory species in which single pair mating are made, the appropriate model will be:

$X_{ij} = \mu + (S + D)_i + e_{ij}$

Where, μ = population mean

$(S+D)_i$ = Effect common to the ith family (F_i)

e_{ij} = random effect specific to the jth full sib in this family.

The covariance between two full sibs would be

$$\text{Cov}(X_{ij}, X_{i1}) = \text{Cov}((\mu + F_i + e_{ij}), (\mu + F_i + e_{ij})) \quad (15)$$
$$= \text{Cov } F_i, F_i$$
$$= \sigma_f^2$$

Heritability is then estimated as:

$$h^2 = \frac{2\sigma_f^2}{\sigma_f^2 + \sigma_e^2}$$

$$= \frac{2\left(\frac{1}{4}\sigma_A^2 + \frac{1}{4}\sigma_{AA}^2 + \frac{1}{4}\sigma_D^2 + \frac{1}{8}\sigma_{AD}^2 \ldots\right)}{P^2}$$

$$= \frac{\sigma_A^2}{\sigma_P^2} \frac{\left(\frac{1}{2}\sigma_{AA}^2 + \frac{1}{2}\sigma_D^2 + \frac{1}{4}\sigma_{AD}^2 + \ldots\right)}{\sigma_P^2} \tag{16}$$

Hare a large potential bias in estimating h² includes dominance and epistatic components of variance. The estimation also depend on the assumptions that:

- The family effect is the sum of the sire and dam genetic effects. This ignores maternal influence and sire and dam interactions.
- There are no environmental covariances between full-sibs.
- Data of plant populations in which each sire and dam are mated to several other individuals e.g. if all dams are mated to all sires, the data are said to form a diallel cross experiment. The appropriate model would be

$$X_{ijk} = \mu + S_i + D_j + (SD)_{ij} + e_{ijk}$$

and estimate of sire and dam variances can be obtained separately from any interaction variance (Becker, 1967).

- Data where both sire and dam are identified, a hierarchical mating arrangement is more useful where each sire is mated to several dams, each of which may have one progeny (in cattle) or more (in pigs, poultry, and sheep).

The model used is then

$$X_{ijk} = \mu + S_i + D_{ij} + e_{ijk}$$

The covariance between two full-sibs is:

Cov (X_{ijk}, X_{ij1}) = Cov $((\mu + S_i + D_{ij} + e_{ijk}), (\mu + S_i + D_{ij} + e_{ij1}))$
= Cov $((S_i + D_{ij}), (S_i + D_{ij}))$

The covariance of two half-sibs for this model is:

Cov (X_{ijk}, X_{im1}) = Cov $((\mu + S_i + D_{ij} + e_{ijk}), (\mu + S_i + D_{im} + e_{im1}))$ (17)

Heritability – Methods of Estimation and Use

= Cov (S_i, S_i)

Thus the dam variance component is equal to the difference between the full sib and half-sib covariance i.e.

= Cov (FS) – Cov (HS).

These variance components are estimated form the analysis of variance as follows:

Source	d.f.	Mean square	Expected mean square
Among sires	s-1	MSS	$\sigma_e^2 + t\sigma_D^2 + td\sigma_S^2$
Among dams within sires	s(d-1)	MSD	$\sigma_e^2 + t\sigma_D^2$
Among progeny within dams	sd(t-1)	MSE	σ_e^2

Then the estimates are;

$$\sigma_e^2 = MSE$$

$$\sigma_d^2 = (MSD - MSE)/nt$$

$$\sigma_S^2 = (MSS - MSD)/td$$

$$\sigma_P^2 = \sigma_S^2 + \sigma_d^2 + \sigma_e^2$$

These estimates provide three estimates of heritability

From the sire variance

$$h^2 \frac{\sigma_s^2}{\sigma_p^2}$$

Form the dam variance

$$h^2 \frac{4\sigma_d^2}{\sigma_p^2}$$

$$= \frac{4((Cov \ FS - Cov \ HS))}{\sigma_P^2}$$

$$= \frac{4\left(\frac{1}{2}\sigma_A^2 + \frac{1}{4}\sigma_{AA}^2 + \frac{1}{4}\sigma_D^2 + \frac{1}{8}\sigma_{AD}^2 + \ldots\ldots\right) - \left(\frac{1}{4}\sigma_A^2 + \frac{1}{16}\sigma_{AA}^2 + \frac{1}{64}\sigma_{AAA}^2 + \ldots\ldots\right)}{\sigma_P^2} \quad (18)$$

$$= \frac{4\left(\frac{1}{2}\sigma_A^2 + \frac{3}{16}\sigma_{AA}^2 + \frac{1}{4}\sigma_D^2 + \frac{1}{8}\sigma_{AD}^2 + \ldots\ldots\right)}{\sigma_P^2}$$

$$= \frac{\sigma_A^2}{\sigma_P^2} \frac{\left(\frac{3}{4}\sigma_{AA}^2 + \sigma_D^2 + \frac{1}{2}\sigma_{AD}^2\right)}{\sigma_P^2}$$

Here the bias in estimation includes the full effect of the dominance variance and $\frac{3}{4}$ of epistatic variance. Maternal effect is also likely to inflate the estimate.

From the sire and dam variance

$$h^2 = \frac{2\left(\sigma_S^2 + \sigma_d^2\right)}{\sigma_P^2}$$

$$= \frac{2(Cov\ FS)}{\sigma_P^2}$$

$$= \frac{2\left(\frac{1}{2}\sigma_A^2 + \frac{1}{4}\sigma_{AA}^2 + \frac{1}{4}\sigma_D^2 + \frac{1}{8}\sigma_{AD}^2 + \ldots\ldots\right)}{\sigma_P^2}$$

$$= \frac{\sigma_A^2}{\sigma_P^2} \frac{\left(\frac{1}{2}\sigma_{AA}^2 + \frac{1}{2}\sigma_D^2 + \frac{1}{4}\sigma_{AD}^2\right)}{\sigma_P^2} \tag{19}$$

Here, the upward bias is intermediate between those of (i) and (ii).

It may be noted here that the difference between the dam and sire variances multiplied by four can be used to estimate the variance due to dominance, in the absence of maternal effects or other non-genetic factors common to a group of full sibs:

$$= 4\left(\sigma_d^2 - \sigma_S^2\right)$$

$$= 4\left(\sigma_d^2 + \sigma_s^2 - 2\sigma_s^2\right)$$

$$= 4 \,(\text{Cov (FS)} - 2 \,\text{Cov (HS)})$$

$$= 4\left(\frac{1}{2}\sigma_A^2 + \frac{1}{4}\sigma_{AA}^2 + \frac{1}{4}\sigma_D^2 + \frac{1}{8}\sigma_{AD}^2 + \ldots\right) - \left(\frac{1}{2}\sigma_A^2 + \frac{1}{8}\sigma_{AA}^2 + \ldots\right)$$
(20)

$$= \sigma_D^2 + \frac{1}{4}\sigma_{AD}^2 + \frac{1}{2}\sigma_{AA}^2 + \ldots$$

The bias in this estimate of dominance variance are the variances due to epistatic effects.

$$a = \sqrt{\frac{2\sigma_d^2}{\sigma_A^2}}$$

A measure of the degree of dominace has been proposed by Comstock and Robinson (1948) as the ratio of under root of twice the dominance variance to the additive variance:

Using $\sigma_D^2 = 4\left(\sigma_d^2 - \sigma_s^2\right)$ and $\sigma_A^2 = 4\sigma_s^2$ we have

$$= \sqrt{\frac{2\left(4\left(\sigma_d^2 - \sigma_s^2\right)\right)}{4\sigma_s^2}}$$

$$= \sqrt{\frac{2\left(\sigma_d^2 - \sigma_s^2\right)}{\sigma_s^2}} \tag{20}$$

Note that the σ_D^2 is for dominace variance and σ_d^2 for dam component of variance.

It often happens that the data for heritability estimation consists of a mixture of half and full-sibs in which only one parent is identifiable. In such cases if heritability is to be estimated for a trait not much influenced by maternal effects, the sire variance component may be used to estimate the additive genetic variance. However, the sire variance is not multiplied by 4 (appropriate if all progeny of a sire were half-sibs) nor hy 2 (appropriate if all progeny of a sire were full-sibs) but by a factor which depend on the ratio of full to half-sibs present. If a sire group consists of 'n' full-sib groups of 'k' individuals each, then the sire variance is made up from a proportion $\frac{(k-1)}{(nk-1)}$ of full sib differences, and a proportion $\frac{k(n-1)}{(nk-1)}$ of half-sib differences. We know that sire variance attributable to the proportion of full-sib differences consists of half the additive genetic variance and that of half sibs contains one fourth of additive genetic variance. Thus the above proportion of full-sibs and half-sibs are multiplied by 0.5 and 0.25 respectively, added and then inversed to obtain the factor by which the sire variance component is multiplied.

$$\frac{k-1}{nk-1}(0.5) + \frac{k(n-1)}{nk-1}(0.25)$$

$$= \frac{k-1}{2(nk-1)} + \frac{k(n-1)}{4(nk-1)}$$

$$= \frac{2(k-1) + k(n-1)}{4(nk-1)}$$

$$= \frac{2k - 2 + nk - k}{4(nk-1)}$$

$$= \frac{nk + k - 2}{4(nk-1)} = \frac{k(n+1) - 2}{4(nk-1)} \qquad (21)$$

The inverse of this is 4 (nk - 1) / k (n+1) - 2.

In case the sires or dams whose variance components are used to estimate heritability, are themselves inbred, then their half or full-sib progeny will have more than the normal genetic covariance. This can be partly compensated for by changing the factor by which sire or dam variance component is multiplied for 4 to $\frac{4}{1+F}$, where 'F' is the inbreeding coefficient of the sire or dam in question.

3. Response of Selection

It was stated earlier that we expect only the fraction of the

Heritability – Methods of Estimation and Use

phenotypic superiority of parents due to additive genetic effects to be transmitted to their offspring. This is shown as follows:

$$\Delta u = h^2 \left(P_{\overline{SD}} - \overline{P} \right)$$
(23)

Where Δu is the expected superiority of the offspring (expected genetic gain), h^2 is heritability in the narrow sense, or the regression of the additive genetic values on the phenotypes, $\left(P_{\overline{SD}} \right)$ is the mean phenotype of the male and female parents and $\left(\overline{P} \right)$ is the population average. Obviously $\left(P_{\overline{SD}} - \overline{P} \right)$ is the phenotypic superiority of the parents (selection differential). If the trait is sex limited or is measured on only one parent (P_D), then

$$\Delta u = \frac{h^2}{2} \left(P_D - \overline{P} \right)$$
(24)

Form equations (11) and (12), it should be obvious that we can now estimate h^2 as follows:

$$h^2 = \frac{\Delta u}{\left(P_{\overline{SD}} - \overline{P} \right)} \quad (22)$$

or $$h^2 = \frac{2 u}{\left(P_D - \overline{P} \right)} \quad (23)$$

If the trait is sex limited. Thus, we may estimate h^2 from date of a selection experiment by determining the phenotypic superiority of the parents (the selection differential) and the response to selection (Ä u). These variables are then substituted into either equation (13) or (14), whichever is appropriate, to estimate h^2. Estimates of heritability obtained in this way are estimates of heritability in the narrow sense, $h^2_{(N)}$.

4. Synthetic Selection

Data from unselected herds may also be used to estimate h^2 by "synthetic selection". We may divide a herd into two groups on the basis of production. We may consider this as "selection" in two directions. The total selection differential would then be equal to the difference in the mean production of the high $(\bar{P}_H)$ and low parental groups $(\bar{P}_L)$ as shown below:

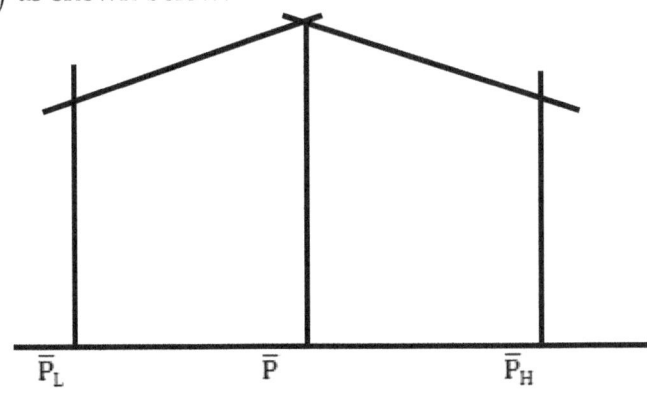

Selection Differential

$$\text{Selection Differential} = (\bar{P}_H - \bar{P}) + (\bar{P} - \bar{P}_L) = \bar{P}_H - \bar{P}_L \quad (24)$$

The total response to selection is equal to the difference in the mean production of the offspring of the high group $(\bar{O}_H)$ and the offspring of the low group $(\bar{O}_L)$.

$$\Delta u = (\bar{O}_H - \bar{O}) + (\bar{O} - \bar{O}_L) = \bar{O}_H - \bar{O}_L \quad (25)$$

Let us consider the following hypothetical example concerned with size of first litter of dams and daughters in a swine herd:

12	)		7	)
10	)	$\bar{P}_H = 9.5$	9	) $\bar{O}_H = 7.5$
9	)		6	)
7	)		6	)
6	)		5	)
5	)		7	)
4	)	$\bar{P}_L = 4.5$	8	) $\bar{O}_L = 6.5$
3	)		6	)

From equation (23),

$$h^2 = \frac{2\Delta u}{\text{Selection Differential}} = \frac{2(7.0-6.5)}{9.5-4.5} = 0.20$$

If data regarding a sex limited trait were available from several herds, h² would be calculated on an intra-herd basis as follows:

$$h^2 = \frac{2(\Delta u_1 + \Delta u_2)}{((Sel.\ Diff.)_1 + (Sel.\ Diff.)_2)} = \frac{2((\overline{O}_{H_1} - \overline{O}_{L_1}) + (\overline{O}_{H_2} - \overline{O}_{L_2}))}{(\overline{P}_{H_1} - \overline{P}_{L_1}) + (\overline{P}_{H_2} - \overline{P}_{L_2})}$$

(29)

The ratio shown in equation (29) should not be doubled when data are available on both sexes and the mid-parent values are considered. Estimates obtained by synthetic selection are estimates of heritability in the narrow sense, $h^2_{(N)}$.

5. Correlation of Identical Twine

The power of twin designs arises from the fact that twins may be either monozygotic (identical (MZ): developing from a single fertilized egg and therefore sharing all of their alleles) – or dizygotic (DZ: developing from two fertilized eggs and therefore sharing on average 50% of their polymorphic alleles, the same level of genetic similarity as found in non-twin siblings). These known differences in genetic similarity, together with a testable assumption of equal environments for identical and fraternal twins[6] creates the basis for the twin design for exploring the effects of genetic and environmental variance on a phenotype.

The basic logic of the twin study can be understood with very little mathematics beyond an understanding of correlation and the concept of variance.

Like all behavior genetic research, the classic twin study begins from assessing the variance of a behavior (called a phenotype by geneticists) in a large group, and attempts to estimate how much of this is due to:

Genetic effects (heritability)

Shared environment - events that happen to both twins, affecting them in the same way;

Unshared, or unique, environment - events that occur to one twin but not another, or events that affect each twin in a different way.

Typically these three components are called A (additive genetics), C (common environment) and E (unique environment); hence the acronym "ACE". It is also possible to examine non-additive genetics effects (often denoted D for dominance (ADE model); see below for more complex twin designs).

The ACE model indicates what proportion of variance in a trait is heritable, versus the proportions which are due to shared environment or unshared environment. Research is carried out using SEM programs such as OpenMx, however the core logic of the twin design is the same, as described below:

Monozygotic (identical - MZ) twins raised in a family share both 100% of their genes, and all of the shared environment. Any differences arising between them in these circumstances are random (unique). The correlation between identical twins provides an estimate of A + C. Dizygotic (DZ) twins also share C, but share on average 50% of their genes: so the correlation between fraternal twins is a direct estimate of ½A+C. If r is correlation, then r_{mz} and r_{dz} are simply the correlations of the trait in identical and fraternal twins respectively. For any particular trait, then:

$$r_{mz} = A + C \tag{26}$$

$$r_{dz} = \tfrac{1}{2}A + C \tag{27}$$

A, therefore, is twice the difference between identical and fraternal twin correlations : the additive genetic effect (Falconer's formula). C is simply the MZ correlation minus this estimate of A. The random (unique) factor E is 1- MZ_r: i.e., MZ twins differ due to unique environments only. (Jinks & Fulker, 1970; Plomin, DeFries, McClearn, & McGuffin, 2001).Stated again, the difference between these two sums, then, allows us to solve for A, C, and E. As the difference between the identical and fraternal correlations is due entirely to a halving of the genetic similarity, the additive genetic effect 'A' is simply twice the difference between the identical and fraternal correlations:

$$A = 2\,(r_{mz} - r_{dz}) \tag{28}$$

As the identical correlation reflects the full effect of A and C, E can be estimated by subtracting this correlation from 1

$$E = 1 - r_{mz}$$

Finally, C can be derived:

$$C = r_{mz} - A$$

Thus, the estimate obtained by using identical twins is even larger than heritability in the broad sense, $h^2_{(B)}$.

Precision of estimates of heritability

While planning an experiment, the experimenter wants to choose those estimates of heritability which will have the greatest precision for the same total number of individuals measured. Thus with a fixed total number of individuals measured, the problem is to balance between large families and many families to minimize the sampling variance.

Parent offspring regression estimate

The variance of the estimated regression coefficient of X on Y (b_{XY}) is

$$\sigma^2_{b_{XY}} = \frac{\sigma^2_X(1-r^2_{XY})}{(n-2)\sigma^2_Y}$$

Where N = the number of pairs of X'S and Y'S so that total number T = 2 N.

The variance of a heritability estimate based on a regression coefficient is simply a function of regression variance. If 'X' is an offspring phenotype and 'Y' is a parental phenotype and $h^2 = 2 b_{XY}$, then the variance of the heritability estimate is

$$\sigma^2_{h^2} = V(2b_{XY})$$

$$= 4\sigma^2_{b_{XY}}$$

$$= \frac{\sigma^2_X(1-r^2_{XY})}{(N-2)\sigma^2_Y}$$

$$= \frac{4}{(N-2)} \left(\frac{\sigma^2_X - \sigma \frac{2}{X} r^2_{XY}}{\sigma^2_Y} \right)$$

$$= \frac{4}{(N-2)} \left(\frac{\sigma^2_X}{\sigma^2_Y} - \frac{\sigma^2_X}{\sigma^2_Y} \cdot \frac{(\sigma_{XY})^2}{\sigma^2_X \cdot \sigma^2_Y} \right)$$

$$= \frac{4}{(N-2)} \left(\frac{\sigma^2_X}{\sigma^2_Y} - b^2_{XY} \right)$$

For large N, -2 can be ignored and with small 'b' $\sigma^2_{b_{XY}}$ can also be ignored so that

$$\sigma^2_{h^2} \frac{4}{N} \frac{\sigma^2_X}{\sigma^2_Y}$$

If it is further assumed that parents and offsprings have equal variances, we have

$$\sigma^2_{h^2} = \frac{4}{N} = \frac{4}{T/2} = \frac{8}{T}$$

It is then possible to decide, for a fixed total number of individuals measured (T), the number of parents (number of families) and numbers of offspring per parent (family size) that will minimize the sampling variance of the heritability. In general, with higher expected heritabilityes fewer offspring per parent and more parents should be measured.

Half and full-sib covariance estimates

Two main methods are used for calculating the precision of such heritability estimates.

The first method uses the fact that these heritability estimates are multiples of the intraclass correlation. For example in a one way classification of sires and progeny within sires, the estimate would be

$$\hat{h}^2 = 4 \frac{\sigma^2_S}{(\sigma^2_S + \sigma^2_C)} = 4\hat{t}$$

and its variance is simply

$$\sigma^2_{h^2} = 16\sigma^2_t$$

$$= 16 \left(\frac{2(1+(n-1t)^2(1-t)^2)}{n(n-1)(N-1)} \right)$$

Where,

N= number of sires

n = number of progeny per sire

Some general conclusions about the efficiency of different

combinations of numbers of progeny per dam, dams per sire and numbers of sires were drawn by Robertson (1959 a) which are as follows:

- Family sizes of the order of 2 to 3 are extremely inefficient
- With a single classification and correlation 't', the optimum family size is in the neighborhood of $1/t$ or $4/h^2$
- With a double classification of sire and dam and equal intraclass correlations in the two cases:
- The best estimate of sire correlation is with one progeny per dam and sire family size as $1/t$.
- For optimum equal information on both sire and dam correlation, the structure has 3 or 4 dams per sire and $\dfrac{1}{2t}$ offspring per dam.
- For a given number of animals measured, the estimate of heritability obtained from a half-sib analysis with optimal structure is more accurate than that form a parent offspring regression if the heritability is less than 0.25 and vice-versa.
- The second measure of precision of heritability, when the variance of a heritability estimate loses much of its value as a measure of precision due to the distribution of the heritability itself being much skewed, is based on the 'F' distribution which overcomes this difficulty and provides a direct confidence interval for the estimate.

The method is as follows for the case of a one-way classification.

ANOVA

Source	d.f.	M.S.	Expected MS	F
Among sires	d_1	M_1	$E_1 = \sigma_e^2 + k\sigma_s^2$	$F_x = \dfrac{M_1}{M_2}$
Among progeny within sires	d_2	M_2	$E_2 = \sigma_e^2$	

The estimate of heritability is;

$$\hat{h}^2 = \dfrac{\sigma_s^2}{(\sigma_s^2 + \sigma_e^2)} = \dfrac{\dfrac{4(E_1 - E_2)}{k}}{\dfrac{(E_1 - E_2)}{k} + E_2} = \dfrac{\dfrac{4(E_1 - E_2)}{k}}{\dfrac{E_1 - E_2 + k E_2}{k}}$$

$$= \frac{4(E_1 - E_2)}{E_1 + (k-1)E_2} = \frac{4\left(\frac{E_1}{E_2} - 1\right)}{\frac{E_1}{E_2} + k - 1} = \frac{4}{1 + \frac{K}{\frac{E_1}{E_2} - 1}}$$

If each sire has an equal number of progeny $\frac{M_1}{M_2}, \frac{E_1}{E_2}$ is distributed F_{d_1}, d_2 (Graybill, 1961) and an exact (1-2á) confidence interval on has lower and upper limits and where F_C is the calculated F from the analysis of variance and $F_á$ and $F_{1-á}$ are the tabulated values for degrees of freedom d_1 and d_2.

This gives directly the following confidence interval for heritability 1 - 2á

Which is exact for equal sized progeny groups, and is otherwise approximate.

Example:

Source	d.f.	M.S.	F
Among sires	30	0.01872	1.707
Among progeny within sires	499	0.01097	

With K = 9.561, $\sigma_s^2 = 0.00081$ $h^2 = 0.275$

To place a 90% confidence interval on heritability, we obtain $F_{.05; 30; 499} = 1.482$ and $F_{.95; 30; 499} = 0.611$ and insert the values

$$\frac{F_C}{F_\alpha} = \frac{1.707}{1.482} = 1.152 \text{ and}$$

$$\frac{F_C}{F_\alpha - 1} = \frac{1.707}{0.611} = 2.794$$

into the formula given above.

The resulting confidence interval is

P { 0.063 d" h^2 d" 0.632} = 0.9

Use of Heritability

In this section, we are concerned with heritability in the narrow sense. Heritability is useful for predicting the response to selection. The expected breeding value of additive genetic value of an animal may be predicted from its phenotype as follows:

$$G = \bar{G} + b_{GP}(P - \bar{P}) = \bar{P} + h^2(P - \bar{P})$$

Where G is the expected breeding value, is assumed to equal and where b_{GP} is equal to $h^2_{(N)}$. Since an individual transmits half of its genetic material to its offspring, it is expected to transmit one-half of its superiority regarding its expected breeding value.

$$G_O = \bar{P} + \frac{h^2}{2}(P - \bar{P}) = \bar{P} + \Delta u$$

Where G_O is the expected breeding value of the offspring, P is the phenotype of one parent and Ä u is the expected genetic gain. If data are available on both parents, then

$$G_O = \bar{P} + h^2(P_{\overline{SD}} - \bar{P}) = \bar{P} + \Delta u$$

For example, if a cow has produced 600 pounds of butter fat and the herd average is 500 pounds, assuming that she is mated to a randomly chosen sire, what is the expected breeding value of her daughter? If $h^2 = 0.25$, then

$G_{Daughter} = 500 + (½)(0.25)(600 - 500) = 512.5$ pound

This estimate may not be very reliable when only one individual is concerned. If, however, a large number of cows which averaged 600 pounds were selected to produce offspring for the next generation, our estimate of 512.5 pounds for the average production of the next generation would be more accurate.

Let us consider another example: a selected group of rams has an average fleece weight of eight pounds and a selected group of ewes has an average fleece weight of six pounds. If the mean fleece weight of the flock is four pounds and if $h^2 = .4$, what is the expected mean fleece weight of the offspring of these selected sheep? From the following equation

$G_O = 4 + (.4)(P_{SD} - 4)$

$P_{\overline{SD}} = \frac{8 + 6}{2} = 7$

$G_O = 4 + (.4)(7 - 4) = 5.2$ pounds

The selection differential, $(P_{SD} - \bar{P})$ or $(P - \bar{P})$, is a function of the selection intensity, i.e. the fraction of the population saved (b):

$$\text{Selection differential} = \left(\frac{Z}{b}\right)(\sigma_p)$$

Where Z is the height of the line at the point of truncation with the normal curve and σ_p is the phenotypic standard deviation. This is illustrated as follows:

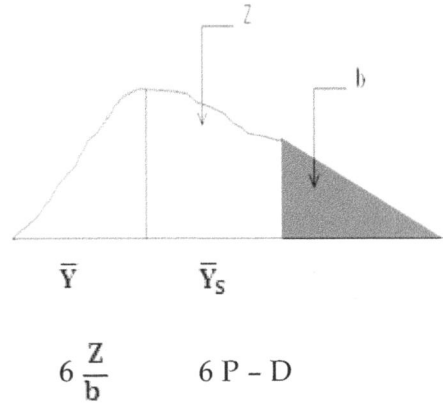

$\frac{Z}{b}$ is the number of standard deviations that the mean of the selected population (Y_S) exceeds the mean of the original population (Y).

The mean of a selected population (Y_S) may be predicted if we know the original population mean and the phenotypic standard deviation.

$$Y_S = Y + \frac{Z}{b}\bar{\sigma}_R$$

For example, the mean of the top one half of a normal population is 0.80 $\bar{\sigma}_p$ above the original population. If = 80 and, then = 500 + (0.8)(80) = 564.

Conversely, if we know the mean of a selected group, we can predict the population mean as follows:

$$\bar{Y} = \bar{Y} - \frac{Z}{b}\sigma_p$$

For example, assume a dairyman who has 100 cows has told you his top ten cows produced 640 pounds of butter fat. You can then predict the mean herd production as follows:

$$\bar{Y} = 640 - \frac{Z}{b} \sigma_P$$

The phenotypic standard deviation of butter fat production is 80 pounds and b = 10%, and $\frac{Z}{b} = 1.75$, therefore,

$$\bar{Y} = 640 - (1.75)(80) = 640 - 140 = 500 \text{ pounds.}$$

This selection differential may also be used for predicting the response to selection.

$$G_O = \bar{P} + \frac{h^2}{Z} \frac{Z}{b} \sigma_P = \bar{P} + \Delta u$$

Where b = % of the population selected to produce offspring for the next generation and where the trait is sex limited. But,

$$G_O = \bar{P} \ h^2 \ \frac{Z}{b} \sigma_P = \bar{P} + \Delta u$$

Where the trait is measured on both males and females. If the selection intensity is higher for one sex than the other, then it is necessary to compute average selection intensity. For example, if one needs to save 20% of the males in his flock to maintain numbers and 50% of his ewes, then $b_S = .20$ and $\frac{Z}{b_s} = 1.4 \ \sigma_P$, where $b_D = .50$ and $\frac{Z}{b_D} = 0.8$ The average selection differential is then

$$\frac{1.4 \ \sigma_P + 0.8 \ \sigma_P}{2} = 1.10 \ \sigma_P$$

Therefore, from equation (18), the expected response to selection for fleece weight is as follows:

$$G_O = \bar{P} + h^2 (1.10) \sigma_P$$

If pounds of fleece, $h^2 = 0.4$ and $\sigma_P = 2$ pounds, then
$G_O = 4 + .4 (1.10)(2) = 4.88$ pounds.

Importance of Heritability

1. Heritability provides a measure of genetic variation, that is, the variation upon which all the possibilities of changing the genetic composition of the population through selection

depend. In other words, knowledge of its magnitude gives the idea about the scope for effecting genetic improvement through selection.

2. It also gives a measure of the accuracy with which the selection for a genotype can be made from a phenotype of the individual or a group of individuals. In individual selection, in which members of the population are selected on the basis of their phenotypic values, the accuracy of selection measured in terms of the correlation between genic values (breeding values), and phenotypic values.

3. Thus the square root of the heritability expresses the reliability of the phenotypic value as a guide to the breeding value.

4. Another important function of heritability is its role in predicting the breeding value of an individual as well as in predicting the genetic improvement expected as a result of the adoption of particular scheme of selection.

5. Thus the best estimate of an individual's breeding value is the product of its phenotypic value and the heritability. The magnitude of heritability dictates the choice of selection method and breeding system.

6. High heritability estimates indicate that additive gene action is more important for that trait, and selective breeding i.e. mating of the best to the best should produce more desirable progeny. Low estimates, on the other hand, indicate that probably non-additive gene action such as overdominance, dominance, and epistacy is important.

There are five important attributes about estimates of heritability and environmentability. They are:

1. Heritability and environmentability are abstract concepts. No matter what the numbers are, heritability estimates tell us nothing about the specific genes that contribute to a trait. Similarly, a numerical estimate of environmentability provides no information about the important environmental variables that influence a behavior.

2. Heritability and environmentability are population concepts. They tell us nothing about an individual. A heritability of .40 informs us that, on average, about 40% of the individual differences that we observe in, say, shyness may in some way

be attributable to genetic individual difference. It does NOT mean that 40% of any person's shyness is due to his/her genes and the other 60% is due to his/her environment.

3. Heritability depends on the range of typical environments in the population that is studied. If the environment of the population is fairly uniform, then heritability may be high, but if the range of environmental differences is very large, then heritability may be low. In different words, if everyone is treated the same environmentally, then any differences that we observe will largely be due to genes; heritability will be large in this case. However, if the environment treats people very differently, then heritability may be small.

4. Environmentability depends on the range of genotypes in the population studied. This is the converse of the point made above. However, it probably does not apply strongly to human behavior as it does to the behavior of specially bred animals. Few — if any — human populations are as genetically homogeneous as breeds of dogs, sheep, etc.

5. Heritability is no cause for therapeutic nihilism. Because heritability depends on the range of typical environments in the population studied, it tells us little about the extreme environmental interventions utilized in some therapies.

Repeatability - Methods of Estimation and Uses

There are two assumptions necessary involved in the regression of repeatability. These are :

- The variances of the different measurements are equal and have their components in the same proportions, and
- The different measurements reflect genetically the same character. If these assumptions are not valid the repeatability concept becomes vague without precise meaning in relation to the variance components.

The repeatability differs very much depending on the following:

- The nature of the trait.
- The genetic properties of the population.
- Environmental conditions under which the individuals are kept or raised.

Repeatability indicates the proportion of observed differences in performance between animals caused by differences in real producing ability. That is say, repeatability of butterfat yield is 0.40, and if a cow has an age corrected butter fat yield of 30 kilograms above the herd average, her near producing ability will be equal to 30x 0.40 = 12 kg. above the herd average. This can also be explained in a different way, i.e., if two cows differ in one lactation by 30 kg in butter fat yield. They will have difference of 12 kg of butter fat in the next lactation.

Repeatability has a value 0 to 1 and may also be expressed in percentage. Repeatability of a trait is not a constant, and measurements error or generally, varying environmental conditions tend to increase V_{Et} and thus to decrease repeatability, shrode et al (1960) have shown the effect of a change in environmental conditions on repeatability. They obtained repeatabilities of milk yield, butter fat yield and fat percentage as 0.37, 0.32 and 0.7 respectively and after improving the management removed some of the environmental variations that had marked the cows real producing abilities.

Table : Repeatability coefficients (r) of the various traits of different livestock species

Trait	Rage of R
1. Lactation milk yield	0.30 – 0.50
2. Lactation fat yield	0.35 – 0.45
3. Fat content	0.50 – 0.70
4. Persistency	0.10 – 0.20
5. Dry period	0.15 – 0.25
6. Calving interval	0.01 – 0.05
7. Service per conception	0.01 – 0.05
8. Gestation length	0.15 – 0.20
9. Weaning weight	0.30 – 0.50
10. Litter size (in cows)	0.12 – 0.18

Ranges of values given are based on majority of published information :

Repeatability of a dairy trait when estimated for a district or region having many dairy herds, it (r) will be larger than that of repeatability found in a herd. This is because herd differences will be included with differences in individual producing ability. Statistically the variance component for herds (VH) occurs both in the numerator and denominator i.e.

$$r = \frac{V_G + V_{EP} + V_H}{V_G + V_{EP} + V_{Et} + V_H}$$

This shows how differences between herds or groups of animals in different environments may irrelevantly increase the repeatability. Repeatability coefficients estimated for a trait from contiguous periods or stage usually tend to be higher than those based on observations spread further apart. Such as Rendal et al (1957) found correlations of first lactation milk yield with 2nd, 3rd and 4th lactations milk yields as $r_{1,2} = 0.50$, n, 3 = 0.43, $r_{1,4} = 0.40$ respectively. On the other hand, there are compensatory influences which may decrease the correlation between contiguous periods or performances. Such as there is negative relationship between dry period and current lactation milk yield (Johanson and Hansson, 1940). However, a longer dry period has beneficial effect on the yield of the following lactation.

Uses of Repeatability

Knowledge of the repeatability of a trait is useful to the animal breeder in following ways.

(i) Uppar limit of heritability

It sets upper limits of heritability (VA/VP) and degree of genetic determination (VG/VP) of the trait. Repeatability is usually easier to determine than either of the above mentioned two ratios.

(ii) Gain from multiple measurements

Repeatability indicates the gain in accuracy expected form multiple measurements. To understand this point assume that each individual is measured n times, and the mean of these n measurements is taken as individual phenotypic value (Pn). Then the VP is made up of the V_G, V_{EP} and the one n^{th} of V_{Et} that is

$$VP_{(n)} = V_G + V_{EP} = \frac{1}{n} V_{Et}$$

Thus, with increase in the number of measurements the variance caused by temporary environmental influences (i.e. V_{Et}) decreases in V_{Et} will be equal to V_{Et}/n. The regression coefficient of the performance potential on averages of n observations becomes.

$$b = \frac{V_G + V_{EP}}{V_G + V_{EP} + \frac{V_{ET}}{n}}$$

$$= \frac{r}{r + 1 - \frac{r}{n}}$$

$$= \frac{nr}{1 + nr - r}$$

$$= \frac{nr}{1 + (n-1)r}$$

Here, r is the repeatability based on one record and n is the number of records available and used. This formula is used to estimate repeatability of traits were a larger number of records is involved. When observations or measurements of a trait are repeated on individual, the variance in the denominator decreased and the regression coefficient (b) of the real producing ability on the actual performance increases. This, therefore, leads to obtain more accurate estimates of the real producing ability.

The VP based on the mean of n measurements (i.e. VP(n)) as a proporting of the VP based on one measurements can be expressed as repeatability.

$$VP(n) = V_G + V_{EP} + 1/n \, V_{Et}$$

$$= \left(r + \frac{1-r}{n}\right) VP, \text{ (since } V_{Et}/VP = 1 - r)$$

$$\frac{VP(n)}{VP} = \frac{nr + 1 - r}{n}$$

$$\frac{1 + r(n-1)}{n}$$

This ratio (VP(n)/VP) shows the trend of reduction in phenotypic variance as a resalt of multiple measurements and as per the repeatabilityies of different traits. When the repeatability is high, i.e. there is little V_{ET} multiple measurements give little gain in accuracy.

When the repeatability is low multiple measurements may yield worthwhile gain in accuracy. For example the repeatabilities of milk constituents, like fat content (r=0.50 - 0.70) are fairy high and, therefore, one (say first) lactation performance estimates accurately a cow's production potential and 2nd lactation record will add very little. Similar is the case with the linear body measurements like wither height, girth, length, etc. of cattle which have fairly high repeatabilyies however, for a trait like litter size in swine which has low repeatability, an average of two or three litters indicates the fertility potential of a cow much better than a single litter record. Therefore, in this case it may pay to wait for a II or even III litter before deciding to cull a sow.

It is, however, noted that with increase in number of measurements the gain in accuracy slows down and after a certain level it falls off rapidly. Therefore, it is rarely of any use to make more than two or three measurements of a trait on an individual. Further, one can take or work with the mean or the sum of measurements as in either case the relative magnitudes of the componenets are not affected.

The advantage of gain in accuracy in breeding programme is that there is increased proportion of additive genetic variance. However, this is true only if two assumptions are valid.

The different measurements have equal variances and

The measurements represent same character genetically.

These assumptions do not hold for milk yield of cows in successive lactations (Rendal et al. 1957) where in the proportion of additive genetic variance is actually less for the mean of several lactations than it is for first lactations only.

Prediction of future performance: The prediction of future performance has no genetical logic and based on the partitioning of the variance into components due to permanents and temporary effects (i.e. made by r). Performance of individuals considered in terms of deviations form the population mean. Suppose an individual has performed better first time partly due to the temporary environmental effects which are not carried through to the subsequent or next performance. Therefore, the subsequent or future performance tends to regress towards the population mean. For example, milk yield, of a cow in her first and second lactations. The repeatability, which is the correlation between the two performances of an individual, indicates the accuracy with which the second lactation yield can be predicted form the first one. The prediction is made from the regression coefficient of second on first performance.

If,

X = first performance of individual

$\bar{X}$ = population mean of first performance

Y = second performance

$\bar{Y}$ = population mean of first performance

b = regression coefficient of Y on X.

Then the prediction of second performance can made by

$(Y - \bar{Y}) = b(X - \bar{X})$

The relationship between the regression coefficient and correlation is shown as

Where σ_Y and σ_X are the standard deviations of X and Y respectively.

Most Probable Producing Ability

Life time averages of individuals are important in animal breeding as these help in identifying the individual having ability to repeat a high level of performance over a long period of time. However, these records should be accurate and corrected for all possible environmental factors before individuals in a herd are compared. Lush (1945) suggested a formula for adjusting the records of cows with different numbers of records to the same basis and called it the probable producing ability (PPA) of an individual, which is

$$PPA = \text{Herd average} + \frac{nr}{1 + (n-1)r} (\text{individuals average} - \text{Herd average})$$

This can be used for culling individuals from herd in case there is considerable variation in ages and numbers of records.

Prichner (1969) has given PPA as the real producing ability (RPA) and the same formula in equation:

RPA $= b(O - M) + M$

Which can be rearranged as?

RPA $= bO - bm + m$

$= bO + M - bm$

$= bO + M(1 - b)$

Where M = Herd (Population) Mean

O = Individuals observed performance

$$b = \frac{nr}{1+(n-1)r} = r(n)$$

If b (or r (n)) is high more the animals own performance is of little value in predicting PPA (or RPA). If no observation is available on the animal (b=o), the best estimate of its performance potential is the population mean.

Repeatability Estimation, Statistical Model and Analysis of Variance

When repeated measurements are made on the same trait of an individual, it is possible to estimate the repeatability (r) of the trait. There are two situations.

When there are equal numbers of measurements available per individual (i.e. balanced design) and

Unequal numbers of measurements per individual (i.e. unbalanced design).

Equal numbers of measurements per individual

Statistical model

$Y_{km} = \mu + a_k + e_{km}$

Where,

u = common mean

αk = effect of Kth individual

ekm = environmental deviation of mth measurements within an individual (all effects random, normal independent with expectations equal to zero).

(a) Analysis of variance table

Source of variation	d.f.	Sum of squares	Mean sum of squares	Estimated mean sum of squares
Between individuals	n - 1	SS_B	MSB	$\sigma_w^2 + K_1\sigma_w^2$
Between measurements within individuals	N(M - 1)	SS_W	MSW	σ_w^2

Where,

N = Number of individuals

M = number of measurements per individual (i.e. equal no. for each individual)

$K_1 = M$

(b) Genetic model

Source of variation	VA	VD	VAA	VAD	VDD	VEP	V_{Et}
	1	1	1	1	1	1	0
	0	0	0	0	0	0	$1\sigma_B^2 =$

represents the differences between the individuals and estimates all the genetic variance (VG) and portion of the environmental variance (VEP) peculiar to the individual σ_W^2 represents the difference among measurements within the individual.

Computation – Analysis of variance

Source of variation	d.f.	S.S.	M.S.
Correlation term (CT)	1	$Y^2\cdots/m$	–
Between individuals	N – 1	$\dfrac{\Sigma Y^2 K}{K\ mk} - CT$	$\dfrac{SS_B}{(N-1)} = MS_B$
Between measurements	N (M – 1)	$\dfrac{\Sigma\Sigma\ y^2}{Km\ Km} - \dfrac{\Sigma Y^2 K}{K\ mk}$	$\dfrac{SS_W}{N(M-D)} = MS_W$

Where M_K = number of measurements of Kth individual

$$\sigma_W^2 = MSW$$

$$\sigma_W^2 = \frac{MSB - MSW}{K_1}$$

Therefore, $R = \dfrac{\sigma_B^2}{\sigma_B^2} + \sigma_W^2$

Reliability = standard error of repeatability, i.e.

$$S.E.(r) = \sqrt{\frac{2(1-R)^2[1+(K-1)(r)]^2}{K(K-1)(N-1)}}$$

As seen here, the S.E.(r) is the square root of the sampling variance of the intra class correlation(r). This is an approximate method assuming that the total number of observations (m) is sufficiently large so that r is normally distributed (Fisher, 1954).

(ii) Unequal numbers of measurement per individual

(a) The computational procedure will be little different in this case as compared to the balanced design given earlier. The differences will be in ANOVA, each Y^2K, is divided by M_K (i.e. number of measurements) taken on the Kth individual.

(b) The degrees of freedom for MSW are the total number of measurements minus C, the number of individuals.

The coefficient $K_1 = \dfrac{1}{N-1}\left(M - \dfrac{mk^2}{m}\right)$

And m = total number of measurements

(c) Variance of variance components,

The $\text{var}(\sigma_W^2)$ is same as given for balanced design, while the $\text{var}(\sigma_B^2)$ is calculated as per Searle (1956)

(d) Standard error of repeatability

The details have been given by Swinger et al (1964) this is an approximate method and normality or r and unequal numbers per group is measured.

$$S.E.(r) = \sqrt{\{\,2(m-1)\,(1-r)^2\,1+K_1-1(r)^2 \,/\, K_1^2(m-n)N-1\,\}}$$

CHAPTER - 11

Correlations

In genetic studies it is necessary to distinguish two causes of correlation; is chiefly pleiotropy, though linkage is a cause of transient correlation, particularly in populations derived from crosses between divergent strains. Pleiotropy is simply the property of a gene whereby it affects two or more characters, so that if the gene is segregating it causes simultaneous variation in the characters as affects. For example, genes that increase growth rate increase both stature and weight, so that they tend to cause correlation between these two characters. The degree of correlation arising from pleiotropy expresses the extent to which two characters are influenced by the same genes. But the correlation resulting from pleiotropy is the overall, or net, effect of all the segregating genes that affect both characters. Some genes may increase both characters, while others increase one and reduce the other the former tend to cause a positive correlation the latter a negative one. So pleiotropy does not necessarily cause a detectable correlation. The correlation resulting from environmental causes is the overall effect of all the environmental factors that vary; some may tend to cause a positive correlation, others a negative one.

Genetic Correlation

Hazel (1943) introduced this statistic and is defined as the correlation between the additive genetic values of two traits. When two traits are considered together we have three population parameters:

Additive genotypic variance for trait 1: σ_{1o}^2

Additive genotypic variance for trait 2: $\sigma_{1\tilde{o}}^2$

Covariance between additive genetic value for trait 1 and additive genetic value for trait $2: \sigma_{10, \tilde{1}0}$

The genetic correlation of trait 1 and trait 2 is then

$$r_g = \frac{\sigma_{10,\tilde{1}0}}{\sqrt{\sigma_{10}^2 \cdot \sigma_{\tilde{1}0}^2}}$$

Therefore, the estimation of genetic correlation involves estimation of the variances which has already been described in estimating heritability, and the estimation of covariance by calculating a measure of the resemblance between relatives for the two traits, and equating this measure to its theoretical value.

Similar to the partitioning of genetic variance for trait, an exactly parallel division is carried out on the genetic covariance between two traits. This is illustrated for on-locus model as follows:

Let the genotypes for the two traits be

$A_i A_j = á_i + á_j + (á\,á)_{ij}$ and

$\tilde{A}_i \tilde{A}_j = \tilde{a}_i + \tilde{a}_j + (\tilde{a}\tilde{a})_{ij}$

The covariance between these two genotypes is then.

$E(A_i A_j, \tilde{A}_i \tilde{A}_j) E_{\alpha_i \alpha_i} + E_{\alpha_j \alpha_j} + E(\alpha\alpha)_{ij}(\tilde{\alpha}\tilde{\alpha})_{ij} = \sigma_{10,\tilde{1}0} + \sigma_{01,\tilde{0}1}$

These two elements are the additive $\left((\sigma)_{10, \tilde{1}0}\right)$ components of the genetic covariance between the traits.

For the general n-locus model, the genetic covariance between the traits is

$$\sigma_{g\tilde{g}} = \sum_{i=0}^{n} \sum_{j=0}^{n} \sigma_{ij,\tilde{i}\tilde{j}}$$

$1 d" i + j d" n$

A general expression for the covariance between the genotype of one individual for one trait and that of another individual for the second trait can be developed, which is analogous to that of the formula for the covariance between two individuals for the same trait, by combing the above formula with that for the probability that two individuals have genes identical by descent.

Thus if 'X' is the genotype of individual 'X' for trait 1 and $\tilde{y}$ is the genotype of individual 'Y' for trait 2, and if 'X' and 'Y' have additive and dominance relationships a_{XY} and d_{XY}, then the covariance of X any is, for 'n' loci.

$$\text{Cov}(X,) = \sum_{i=0}^{n} \sum_{j=0}^{n} a_{XY}^{i} d_{XY}^{j} \sigma_{1j,ij}$$

$1 \leq i + j \leq n$

Importance of Genetic Correlation

Genetic correlation gives an idea about the extent to which the two characters are under the control of the same set of genes of have the same physiological basis for their expression. If the correlation is high then pleiotropy is probably more important, if the correlation is low, the two traits are inherited more of less independently or they are under the control of different sets of genes.

Magnitude and sign of genetic correlation help in judging how the improvement in one character will cause simultaneous change in the other character. If the genetic correlation is positive, then improvement of one character by selection automatically results in the improvement in the other, even though direct selection for its improvement has not been made. If the genetic correlation is negative, the improvement of one character through selection will result in a decline in the other e.g. butterfat percentage and milk yield in dairy cattle shows a negative genetic correlation.

Estimation of the Genetic Correlation

Parent offspring regression:

If the phenotype of the parent for trait 1 is $P = G + E$

and that of the offspring for trait 2 is ,

and if the environmental components of the two models are uncorrelated with each other or with the genetic components, then the covariance between parent and offspring phenotypes is

$$\text{Cov}(P, \tilde{P}) = \text{Cov}((G + E), (\tilde{g} + \tilde{e}))$$

$$= \text{Cov}(G, \tilde{g})$$

$$= \frac{1}{2} \sigma_{10,\tilde{10}} + \frac{1}{4} \sigma_{20,\tilde{20}} + \frac{1}{8} \sigma_{30,\tilde{30}} + \ldots$$

$$= \frac{1}{2}\sigma_{A,\tilde{A}} + \frac{1}{2}\sigma_{AA,\widetilde{AA}} + \frac{1}{8}\sigma_{AAA,\widetilde{AAA}} + \ldots$$

Therefore, twice this phenotypic covariance estimates the additive genetic covariance for the two traits plus some epistatic upward bias. The genetic correlation can be estimated, if separate estimates of the additive genetic variances for the two traits are available. However, the two genetic variances should come from the same source, and this would require that phenotypic measures for both traits in parent and offspring be available. When this is so, the four resulting phenotypes are represented as follows:

(Hazel, 1943)

		Trait 1					Trait 2		
Parent	P	=	G	+	E	$\tilde{P}$ =	$\tilde{G}$	+	$\tilde{E}$
Offspring	P	=	g	+	e	$\tilde{p}$ =	$\tilde{g}$	+	$\tilde{e}$

It is assumed that all covariances between genotypic and environmental components are zero. So also, all covariances between environmental components are zero, except for that between E and $\tilde{E}$ and between e and $\tilde{e}$, which will not be zero as they occur in the same individual. For estimating genetic correlation, correlations or regressions between offspring and parental phenotypes could be used to estimate the genetic correlation. However, only the regressions of offspring on parent will be free of bias due to selection among parents, and therefore the four regressions of offspring on parent (indicated by arrows in the above diagram) will be used for estimation of the genetic correlation between trait 1 and 2 as follows:

$$r_g = \sqrt{b_{P_p} b_{\hat{P}_p} \over b_{P_p} b_{\hat{P}_p}}$$

$$= \sqrt{\frac{\text{Cov}(P,\hat{p})}{V(P)} \cdot \frac{\text{Cov}(\hat{P},\hat{p})}{V(\hat{P})} \over \frac{\text{Cov}(P,\hat{p})}{V(P)} \cdot \frac{\text{Cov}(\hat{P},\hat{p})}{V(\hat{P})}}$$

$$= \sqrt{\frac{Cov(P,\hat{p})\ Cov(\hat{P},p)}{cov(P,P)\ Cov(\hat{P},\hat{p})}}$$

$$= \sqrt{\frac{Cov((G+E),(\hat{g}+\hat{e}))\ Cov((\hat{G}+\hat{E}),(g+e))}{Cov((G+E),(g+e))\ Cov((\hat{G}+\hat{E}),(\hat{g}+\hat{e}))}}$$

$$= \sqrt{\frac{Cov(G,\hat{g})\ Cov(\hat{G},g)}{Cov(G,g)\ Cov(\hat{G},\hat{g})}}$$

$$= \sqrt{\frac{\left(\frac{1}{2}\sigma_{10\,\tilde{1}0}+\frac{1}{4}\sigma_{20\,\tilde{2}0}+\cdots\right)^2}{\left(\frac{1}{2}\sigma^2_{10}+\frac{1}{4}\sigma^2_{20}+\cdots\right)\left(\frac{1}{2}\sigma^2_{\tilde{1}0}+\frac{1}{4}\sigma^2_{\tilde{2}0}+\cdots\right)}}$$

Multiplying both numerator & denominator by 2, we get

$$\frac{\sigma_{10\,\tilde{1}0}+\frac{1}{2}\sigma_{20\,\tilde{2}0}+\cdots}{\sqrt{\left(\sigma^2_{10}+\frac{1}{2}\sigma^2_{20}+\cdots\right)\left(\sigma^2_{\tilde{1}0}+\frac{1}{2}\sigma^2_{\tilde{2}0}+\cdots\right)}}$$

Considering additive x additive and higher order epistatic term as zero, we get

$$= \frac{\sigma_{10\,\tilde{1}0}}{\sqrt{\sigma^2_{10}\ \sigma^2_{\tilde{1}0}}}$$

It may be noted that estimate of genetic correlations can be obtained by using the four-parent-offspring covariances as the denominators of the four regressions cancel out. Further, sing Cov $(G,\hat{g})$ and Cov $(\tilde{G},g)$ both estimate $\left(\frac{1}{2}\sigma_{10\,\tilde{1}0}+\frac{1}{4}\sigma_{20\,\tilde{2}0}+\cdots\right)$, their sum estimates the numerator of the genetic correlation. therefore, another estimate of genetic correlation can be derived, using this numerator but using the same denominator as in the formula given above. Since 2 is no longer multiplied with numerator and denominator while estimating numerator by the above additive method, while using the same denominator as above, the denominator must be multiplied by 2 to produce estimates of the required variances σ^2_{10} and $\sigma^2_{\tilde{10}}$. Thus estimate of genetic correlation is then

$$r_g = \frac{Cov(P,\hat{p}) \; Cov(\hat{P}, p)}{\sqrt[2]{Cov(P, p) \; Cov(\hat{P},\hat{p})}}$$

Which gives the same theoretical value as the previous method. For unselected populations, the additive method may be preferable, to the geometric method since additive method will be less subject to sampling errors (Hazel, 1943; Van Vleck and Henderson 1961; Becker, 1967).

Half and full-Sib covariances: In calculating genetic correlations, the additive genetic variance for each trait can be obtained exactly as in the heritability case. The estimate of the additive genetic component of the covariance between the two traits $\sigma_{10,\tilde{10}}$ can be obtained from an analysis of covariance, carried out similar to the analysis of variance.

Let the following models represent the phenotypes for traits 1 and 2 measured on an individual

Trait 1 : $X_{ijk} = \mu + S_i + D_{ij} + e_{ijk}$

Trait 2 : $\tilde{X}_{ijk} = \tilde{\mu} + \tilde{S}_i + \tilde{D}_{ij} + \tilde{e}_{ijk}$

Then the covariance between trait 1 in one full-sib and trait 2 in a second full-sib is

$Cov(X_{ijk}, \tilde{X}_{ij1}) = Cov((\mu + S_i + D_{ij} + e_{ijk}),(\mu + S_i + D_{ij} + e_{ij1}))$

$= Cov \; S_i \tilde{S}_i + Cov \; D_{ij} \tilde{D}_{ij}$

$= \sigma_{S_{12}} + \sigma_{d_{12}}$

The covariance between trait 1 in one half-sib and trait 2 in a second half-sib is

$Cov(X_{ijk}, \tilde{X}_{im1}) = Cov((\mu + S_i + D_{ij} + e_{ijk}),(\mu + \tilde{S}_i + \tilde{D}_{im} + \tilde{e}_{im1}))$

$= Cov \; S_i \; \tilde{S}_i$

$= \sigma_{S_{12}}$

Here and which are statistical components of covariance for sire and dam respectively, can be obtained from an analysis of covariance

Correlations

Source	d.f.	Mean cross products	Expected mean cross product
Among sires	S - 1	MCPS	$\sigma_{e_{12}} + t\sigma_{d_{12}} + td\sigma_{s_{12}}$
Among dams within sires	S(d-1)	MCPD	$\sigma_{e_{12}} + t\sigma_{d_{12}}$
Among progeny within dams	Sd(t-1)	MCPE	$\sigma_{e_{12}}$

The estimates are:

$\hat{\sigma}_{e_{12}} = \text{MCPE}$

$\hat{\sigma}_{d_{12}} = (\text{MCPD} - \text{MCPE}) / t$

$\hat{\sigma}_{s_{12}} = (\text{MCPS} - \text{MCPD}) / td$

Sire, dam and sire + dam components of covariance are used to estimate the genetic correlation as in case of heritability estimation using the corresponding variance components.

Sire component of covariance (HS covariance):

$$\hat{r}_g = \frac{\sigma_{s_{12}}}{\sqrt{\sigma_{s_1}^2 \, \sigma_{s_2}^2}}$$

$$= \frac{\frac{1}{4}\sigma_{10.\tilde{10}} + \frac{1}{16}\sigma_{20.\tilde{20}} + \cdots}{\sqrt{\left(\frac{1}{4}\sigma_{\tilde{10}}^2 + \frac{1}{16}\sigma_{\tilde{20}}^2 + \cdots\right)\left(\frac{1}{4}\sigma_{10}^2 + \frac{1}{16}\sigma_{20}^2 + \cdots\right)}}$$

$$= \frac{\sigma_{10.\tilde{10}} + \frac{1}{4}\sigma_{20.\tilde{20}} + \cdots}{\sqrt{\left(\sigma_{\tilde{10}}^2 + \frac{1}{4}\sigma_{\tilde{20}}^2 + \cdots\right)\left(\sigma_{10}^2 + \frac{1}{4}\sigma_{20}^2 + \cdots\right)}}$$

Here the theoretical bias is less than that in the regression of offspring on parent method.

Dam components of covariance (FS - HS Covariances):

$$\hat{r}_g = \frac{\sigma_{d_{12}}}{\sqrt{\sigma_{d_1}^2 \, \sigma_{d_2}^2}}$$

$$= \frac{\sigma_{10,\tilde{1}0} + \frac{3}{4}\sigma_{20,\tilde{2}0} + \frac{1}{2}\sigma_{01,\tilde{0}1} + \cdots}{\sqrt{\left(\sigma_{10}^2 + \frac{3}{4}\sigma_{20}^2 + \frac{1}{2}\sigma_{01}^2 + \cdots\right)\left(\sigma_{\tilde{1}0}^2 + \frac{3}{4}\sigma_{\tilde{2}0}^2 + \frac{1}{2}\sigma_{\tilde{0}1}^2 + \cdots\right)}}$$

Here the bias is the dominance effects and interactions in each of the 3 components of the estimate.

Sire and dam components of covariance (FS covariances):

$$\hat{r}_g = \frac{\sigma_{s_{12}} + \sigma_{d_{12}}}{\sqrt{\left(\sigma_{s_1}^2 + \sigma_{d_1}^2\right)\left(\sigma_{s_2}^2 + \sigma_{d_2}^2\right)}}$$

$$= \frac{\frac{1}{2}\sigma_{10,\tilde{1}0} + \frac{1}{4}\sigma_{20,\tilde{2}0} + \frac{1}{4}\sigma_{01,\tilde{0}1} + \cdots}{\sqrt{\left(\frac{1}{2}\sigma_{10}^2 + \frac{1}{4}\sigma_{20}^2 + \frac{1}{4}\sigma_{01}^2 + \cdots\right)\left(\frac{1}{2}\sigma_{\tilde{1}0}^2 + \frac{1}{4}\sigma_{\tilde{2}0}^2 + \frac{1}{4}\sigma_{\tilde{0}1}^2 + \cdots\right)}}$$

$$= \frac{\sigma_{10,\tilde{1}0} + \frac{1}{2}\sigma_{20,\tilde{2}0} + \frac{1}{2}\sigma_{01,\tilde{0}1} + \cdots}{\sqrt{\left(\sigma_{10}^2 + \frac{1}{2}\sigma_{20}^2 + \frac{1}{2}\sigma_{01}^2 + \cdots\right)\left(\sigma_{\tilde{1}0}^2 + \frac{1}{2}\sigma_{\tilde{2}0}^2 + \frac{1}{2}\sigma_{\tilde{0}1}^2 + \cdots\right)}}$$

Here the theoretical bias is less than (b).

Phenotypic and Environmental Correlations

The analytical model used for statistical estimation purposes is
$X_{ijk} = \mu + S_i + D_{ij} + e_{ijk}$

Parallel to this, there is a hypothetical general model representing phenotype, genotype and environmental component:

$P = g + E$

The phenotypic variance is

$\sigma_P^2 = \sigma_g^2 + \sigma_E^2$

The covariance between this phenotype, and that for a second trait whose general model is

$\tilde{P} = \tilde{g} + \tilde{E}$

is $\sigma_{P\tilde{P}} = \sigma_{g\tilde{g}} + \sigma_{E\tilde{E}}$

the phenotypic correlation is now defined as

$$r_P = \frac{\sigma_{P\tilde{P}}}{\sqrt{\sigma_P^2 \sigma_{\tilde{P}}^2}}$$

and the environmental correlation as

$$r_e = \frac{\sigma_{EE}}{\sqrt{\sigma_E^2 \; \sigma_{\tilde{E}}^2}}$$

Estimation of Phenotypic Correlation

Since X_{ijk} is a phenotypic observation, say for the first trait, we have

$$\tilde{\sigma}_P^2 = \sigma_{\tilde{X}}^2 = \sigma_{s_1}^2 + \sigma_{d_1}^2 + \sigma_{e_1}^2.$$

For the second trait

$$\tilde{\sigma}_{\tilde{P}}^2 = \sigma_{\tilde{X}}^2 = \sigma_{s_2}^2 + \sigma_{d_2}^2 + \sigma_{e_2}^2$$

The phenotypic covariance is

$$\sigma_{P\tilde{P}} = \sigma_{X\tilde{X}} = \sigma_{s_{12}} + \sigma_{d_{12}} + \sigma_{e_{12}}$$

These three estimates are put together to give

$$r_p = \frac{\sigma_{s_{12}} + \sigma_{d_{12}} + \sigma_{e_{12}}}{\sqrt{(\sigma_{s_1}^2 + \sigma_{d_1}^2 + \sigma_{e_1}^2)(\sigma_{s_2}^2 + \sigma_{d_2}^2 + \sigma_{e_2}^2)}}$$

Estimation of Environmental Correlation

The estimation of environmental correlation is some what are involved. It is to be considered as the correlation between 'E' terms in the general hypothetical model and not between 'k' (error terms) of the analytical model. For the purposes of correlation estimation, the genotypic compment, g, in the general hypothetical modal is usually defined to be the strictly additive genetic effect, and the environment part 'E' then also contains any non-additive genetic effects. Its variance is therefore best estimated by subtracting the additive genetic variance form the phenotypic variance. The phenotypic variance form the analytical model is

$$\sigma_E^2 = \sigma_s^2 + \sigma_d^2 + \sigma_e^2$$

And the additive genetic variance can be estimated with varying degrees of bias as

$$4\sigma_s^2, \quad 4\sigma_d^2 \text{ or } 2(\sigma_s^2 + \sigma_d^2)$$

These give 3 possible estimates of the 'environmental' variance of the general model:

$$\hat{\sigma}_E^2 = \sigma_P^2 - 4\sigma_S^2 = \sigma_d^2 + \sigma_e^2 - 3\sigma_S^2$$
$$\hat{\sigma}_E^2 = \sigma_P^2 - 4\sigma_d^2 = \sigma_S^2 + \sigma_e^2 - 3\sigma_d^2$$
$$\hat{\sigma}_E^2 = \sigma_P^2 - 2(\sigma_S^2 + \sigma_D^2) = \sigma_e^2 - \sigma_S^2 - \sigma_d^2$$

The environmental component of covariance between two traits can also be estimated similarly. Therefore the environmental can be estimated by any of the three following formulae:

$$r_e = \frac{\sigma_{d_{12}} + \sigma_{e_{12}} - \sigma_{s_{12}}}{\sqrt{([\sigma)]_{d_1}^2 + \sigma_{e_1}^2 - 3\sigma_{s_1}^2)[(\sigma)]_{d_2}^2 + \sigma_{e_2}^2 - 3\sigma_{s_2}^2)}}$$

$$r_e = \frac{\sigma_{s_{12}} + \sigma_{e_{12}} - \sigma_{d_{12}}}{\sqrt{([\sigma)]_{s_1}^2 + \sigma_{e_1}^2 - 3\sigma_{d_1}^2)[(\sigma)]_{s_2}^2 + \sigma_{e_2}^2 - 3\sigma_{d_2}^2)}}$$

$$r_e = \frac{\sigma_{s_{12}} + \sigma_{e_{12}} - \sigma_{d_{12}}}{\sqrt{([\sigma)]_{e_1}^2 + \sigma_{s_1}^2 - \sigma_{d_1}^2)[(\sigma)]_{e_2}^2 + \sigma_{s_2}^2 - \sigma_{d_2}^2)}}$$

Since the sire variance component is the least biased estimator of the additive genetic variance, the first is the best of the three formulae used for environmental correlation.

Precision of Estimates of Genetic Correlation

Robertson (1959b) has given Reeve's formulae (1955) of appropriate sampling variances for the geometric and arithmetic combinations of parent offspring regressions used estimators of the genetic correlation, in terms of the genetic correlation and the heritability of the two traits as;

$$\sigma_{r_g}^2 = \frac{(1-[r_g^2])^2}{N-1} \cdot \frac{2}{h_1^2 h_1^2} + \frac{1}{2} \cdot$$

Van Vleck and Henderson (1961) found Reeve's Formulae accurate for large samples (over 1000 parent – offspring pairs), and inaccurate for smaller samples (500 pairs) except where the heritabilities were high. They concluded that the precision of genetic correlations will be low with less than 1000 pairs of observation, especially when traits have low heritability and the arithmetic method is likely to give the best estimate.

Robertson (1959b) found the approximate relationship between the variances of the heritabilities of the two traits and that of the genetic correlation to be

$$\frac{\sigma_{\hat{r}_g}^2}{\sigma_{\hat{h}_g}^2 \sigma_{\hat{h}_g}^2} = \frac{(1 - r_g^2)^2}{2\, h_1^2\, h_2^2}$$

Which held for a reasonable range of conditions. This gives the variance of $\hat{r}_g$ as:

$$\sigma_{\hat{r}_g}^2 = \frac{(1 - r_g^2)^2}{2} \cdot \frac{\sigma_{\hat{h}_1}^2\, \sigma_{\hat{h}_2}^2}{h_1^2\, h_2^2}$$

This value in general will be small when the standard errors of the heritability estimates are small. Therefore, the best combination of sires, dams and progeny for estimating genetic correlations will be similar to that for estimating heritabilities.

A general formula for the variance of correlation (r) is given by Mode and Robinson (1959), Tallis (1959) and Scheinberg (1966) who used it to give the variances of environmental and phenotypic correlations, which is given below:

If any correlation

$$r = \frac{\sigma_{12}}{\sqrt{\sigma_1^2\, \sigma_2^2}}$$

And if the variances of its three components are:

$V(\sigma_1^2), V(\sigma_2^2)$ and $V(\sigma_{12})$ and their covariances are $\mathrm{Cov}(\sigma_{12}, \sigma_1^2), \mathrm{Cov}(\sigma_{12}, \sigma_2^2)$ and $\mathrm{Cov}(\sigma_1^2, \sigma_2^2)$

Then the variance of the correlation (r) is

$$\sigma_{\hat{r}}^2 = r^2 \left[\frac{V(\sigma_{12})}{\sigma_{12}^2} + \frac{V(\sigma_1^2)}{4(\sigma_1^2)^2} + \frac{V(\sigma_2^2)}{4(\sigma_2^2)^2} \right]$$

The three variances and three covariances can be calculated separately, since each is a function of mean squares whose variances and covariances are known.

Relationships among parameters:

The phenotypic, genetic and environmental correlations are

$$r_P = \frac{\sigma_P \tilde{P}}{\sigma_P \sigma_{\tilde{P}}}, \quad r_g = \frac{\sigma_g \tilde{g}}{\sigma_g \sigma_{\tilde{g}}}, \quad r_E = \frac{\sigma_E \tilde{E}}{\sigma_E \sigma_{\tilde{E}}}$$

Since $\sigma_P^2 = \frac{\sigma_g^2}{h_1^2}$ and $\sigma_{\tilde{P}}^2 = \frac{\sigma_{\tilde{g}}^2}{h_1^2}$ (1) because

$$h_1^2 = \frac{\sigma_g^2}{\sigma_P^2} \quad \text{and} \quad h_2^2 = \frac{\sigma_{\tilde{g}}^2}{\sigma_{\tilde{P}}^2}$$

And since $(1 - h^2) = 1 - \frac{\sigma_g^2}{\sigma_P^2} = \frac{\sigma_E^2}{\sigma_P^2}$

$$\sigma_P^2 = \frac{\sigma_E^2}{(1 - h_1^2)} \quad \text{and} \quad \sigma_g^2 = \frac{\sigma_E^2}{(1 - h_2^2)} \quad (2)$$

Furthermore

Using (1), (2) and (3) above, the phenotypic correlation can be written as

$$r_P = \frac{\sigma_g \tilde{g}}{\left(\frac{\sigma_g}{h_1}\right)\left(\frac{\sigma_g}{h_2}\right)} + \frac{\sigma_E \tilde{E}}{\left(\frac{\sigma_E}{\sqrt{1 - h_1^2}}\right)\left(\frac{\sigma_E}{\sqrt{1 - h_2^2}}\right)}$$

$$= h_1 h_2 r_g + \sqrt{(1 - h_1^2)(1 - h_2^2)}\, r_E$$

Using the above relationship Searle (1961) examined the possible combinations of the three correlations and two heritabilities and concluded:

r_e is negative when r_p and r_g have the same sign only if $\frac{r_p}{r_g} < h_1 h_2$; r_E is negative r_p and are of opposite sign and r_g is negative.

Equality of heritabilities implies that when any two of the correlations are equal, there is equality of all three.

Exceeds (or is less than) according as the ratio excepts (or is less than) the value of

$$\frac{(1 - h_1 h_2)}{\sqrt{(1 - h_1^2)(1 - h_2^2)}}$$

Correlations

Example:

The analysis of variance and covariance for two traits are as follows

Source	d.f.	Mean Squares		Mean cross product XY	Expected mean square or cross product
		X	Y		
Sires	48	77.6481	38.6442	35.5160	$\sigma_e^2 + 2\sigma_d^2 + 8\sigma_s^2$
Dams	192	30.6799	11.4386	15.0266	$\sigma_e^2 + 2\sigma_d^2$
Remainder	239	10.0099	4.0039	4.1128	σ_e^2

The variance and covariance components estimated by equating the mean squares and their expected values are

	X	Y	XY
Sire	$\sigma_{s_1}^2 = 5.8704$	$\sigma_{s_2}^2 = 3.4007$	$\sigma_{s_{12}}^2 = 2.5612$
Dam	$\sigma_{d_1}^2 = 10.3350$	$\sigma_{d_2}^2 = 3.7174$	$\sigma_{d_{12}}^2 = 5.4509$
Error	$\sigma_{e_1}^2 = 10.0099$	$\sigma_{e_2}^2 = 4.0039$	$\sigma_{e_{12}}^2 = 4.1128$

Much larger dam than sire variance for trait x and be/ indicative of material effects and/or sub-stantial dominance and dominance related epistatic variance for this trait.

Therefore, heritability is estimated only from the sire component as.

$$\hat{h}_1^2 = \frac{4(5.8704)}{5.8704 + 10.3350 + 10.0099} = 0.9$$

$$\hat{h}_2^2 = \frac{4(3.4007)}{3.4007 + 3.7174 + 4.0039} = 1.2$$

The correlations are estimated as

$$\hat{r}_g = \frac{2.5612}{\sqrt{(5.8704)(3.4007)}} = 0.5732$$

$$\hat{r}_p = \frac{2.5612 + 5.4509 + 4.1128}{\sqrt{(5.8704 + 10.3350 + 10.0099)(3.4007 + 3.7174 + 4.0039)}}$$

$$= \frac{12.1240}{\sqrt{(26.2153)(11.1220)}} = 0.7099$$

r_E can not be calculated because the estimate of the additive genetic variance for one trait 'Y' is greater than that for the phenotypic variance (σ_A^2 = 4 x 3.4007 > 11.1220), implying at best a zero estimate of σ_E^2.

Chapter - 12

Selection – Natural Vs Artificial Selection

Selection is an important tool for changing gene frequencies to better fit individuals for a particular purpose. It may be defined as a process in which certain individuals in a population are preferred to others for the production of the next generation. Selection is of two kinds

i. Natural

ii. Artificial.

Natural Selection

In nature, the main force responsible for selection is of interest because of its apparent effectiveness and because of principles involved. Some of the most interesting cases of natural selection are those involving man himself. All races of man that now exist belong to the same species, because the races are interfile or have been in all instances where mating have been made between them. All races of man now in existence had a common origin and at one time probably all man had the some kind of skim pigmentation – which kind we have no sure way of knowing. As the number of generations of man increased, mutations occurred in the genes affecting pigmentation of the skin causing genetic variations in this trait over a range from light to dark or black.

Natural selection is a very complicated process and many factors determine the proportion of individuals that will reproduce. Among these factors are differences in mortality of the individuals in the population especially early in life, differences in the duration of the period of sexual activity and in the degree of sexual activity itself and the differences in the degree of fertility of individuals in the population.

Artificial Selection

Artificial selection is selection practiced by man. It may be defined as the efforts of man to increase the frequency of desirable genes, or combinations of genes, in his herd or flock by locating and saving for breeding purposes those individuals with superior performance or which have the ability to produce superior performing offspring when mated with individuals from other lines or breeds.

Artificial selection is divided in two kinds:

i. Automatic
ii. Deliberate selection

Litter six in swine may be used as an illustration of the meaning of these two terms. Here, automatic selection would result from differences in litter size even if parents were chosen entirely at random from all individuals available at sexual maturity under these conditions there would be twice as much chance of saving offspring from breeding purpose from a litter of eight than from a litter of four. Automatic selection here differs from natural selection only to the extent that the size of the litter in which an individual is reared influences the natural selective advantage of the individual for other traits.

Deliberate selection – is the term applied to selection in swine for litter size above and beyond that which was automatic? Must of the selection for litter size at both was automatic and very little was deliberate.

Genetic effect of selection - Selection does not create new genes. Selection is practiced to increase the frequency of desirable genes in a population and to decrease the frequency of undesirable genes.

P_1	AA	X	aa
F_1	All	Aa	
F_2	Aa	X	Aa
		1AA	
Progeny		2 Aa	
		1 aa	

Let us assume that we cull all aa individuals in F_2. The remaining genes would be increase.

Selection for Different Kinds of Genes

Both quantitative and qualitative traits may be greatly affected by many different kinds of gene action. It seems important here to outline what methods may be used in selecting for or against these traits.

Selection for dominant gene

In practice, we are varying likely to be selecting for a dominant gene because traits determined by such genes are usually desirable. Those individuals possessing a dominant gene will show it, but the problem here is one of distinguishing between the homozygous dominant and heterozygous dominant individuals. The heterozygous individual must be identifies by a breeding test or a knowledge of the parental phenotype in some case before they can be eliminated. Selection for dominant genes involves the same principles as selection against a recessive gene.

Selection against a dominant gene

Selection against a dominant gene is relatively easy, providing the penetrance of the gene is 100% and it does not vary in its expression since each animal possessing a dominant trait should show this in its phenotype, eliminating the gene merely means that all animals showing the trait should be discarded, whether or not this can be done at once of course, depends upon numbers of animals possessing the trait and whether one can afford to discard all of them at one time.

If the penetrance of the gene is low and the genes are variable in their expression, selection against a dominant gene would be much less effective. Selection for such a trait could not be based upon the individual's phenotype alone, but attention to the phenotype of ancestor's progeny, and collateral relatives would also be necessary if selection is to be successful.

Selection for recessive gene

Selection for a recessive gene is relatively simple if penetrance is complete. If the genes do not vary too much in their expression and the frequency of the recessive gene is relatively high. Selection under such conditions is merely a matter of keeping those individuals which show the recessive trait. A good example of such selection would be for the horned gene in cattle.

Selection against a recessive gene

Selection against a recessive gene is the same as selection for a dominant gene. In both instances the homozygous recessive individuals can be identified and discarded. Even when this is done, the recessive gene still remains in the herd or population being possessed by heterozygous dominant individuals. To eliminate the recessive gene entirely, the homozygous recessive and heterozygous dominant individuals both must be discarded leaving only the homozygous dominant individuals.

The following is a formula for determining the frequency of a gene in a population in which all of the homozygous recessive individuals are discarded. It is assumed that the heterozygous individuals are not favoured in selection.

$$Fn = \frac{Fo}{1+(N \ X \ Fo)}$$

Where,

Fn = Is the frequency of the recessive gene after all homozygous recessive individuals have been discarded from generation?

Fo = Is the original frequency of the recessive gene before the homozygous recessive individuals discarded?

N = Is the number of generations of selection against the homozygous recessive individuals?

Chapter - 13

Response to Selection

The response to selection is the change produced in the population mean through selection. This is achieved through selecting as parents of the next generation, individuals with highest genetic merit and whose progeny will, as a group, have the highest possible genetic merit for the trait in question.

If parental and progeny phenotypes are:

P = G + E and P = g + e,

Then the change in the progeny genotype (g) per unit change in parental phenotype (p) is the regression of 'g' on P.

$$b_{gP} = \frac{Cov(p,q)}{\sigma_P^2} = \frac{Cov(G,q)}{\sigma_P^2}$$

(Assuming genotype and environment independent and therefore their covariance as zero)

Thus the average value of progeny genotypes in response to selection of parental phenotypes depends largely on Cov (G, g), the genetic covariance of parent and offspring which is half the additive genetic variance of the parents.

We know that on an average, half of genes of a parent are passed to the offspring and with them a random half of his additive merit. One-fourth of his additive x additive gene combinations are also transmitted and lesser fractions of higher additive combinations. No dominance combinations are transmitted, since only one member of each allelic gene pair reaches the offspring. Similarly all epistatic gene combinations involving allelic gene pairs are broken up at gamete formation. If dominance and epistatic gene combinations involving allelic gene pairs are important in determining genetic superiority for the trait, then direct selection of superior parents will not produce genetically superior offspring. Some form of mating system which will recreate these combinations in the offspring, together with the formation of lines or groups within the population is needed in these circumstances.

Thus, the aim of selection is to identify and select as parents individuals with high additive genetic merit. By selecting the parent on additive merit, we also ensure that a fraction of any parallel merit for additive type epistasis is also passed on. In considering response to selection, little is lost by treating the response as if it were wholly determined by the additive effects of genes.

Components of The Response

As only phenotypes and not genotypes could be observed, the average superiority of the selected parents, termed as selection differential (s) is measured as the mean phenotypic value of the individuals selected as parents expressed as a deviation from the population mean, i.e. from the mean phenotypic value of all the individuals in the parental generation before selection was made. The difference of mean phenotypic value between the offspring of the selected parents and the whole of the parental generation before selection is called response to selection (R).

The Connection between Response and selection differential can be deduced as under

Two successive generations of a population mating at random, is represented below diagrammatically. Each point represents a pair of parents and their progeny, and is marked according to the mid-parent value measured along the horizontal axis and the mean value of the progeny measured along with vertical axis. The origin represents the population mean, which is assumed to be the same in both generations. The regression line of offspring on mid-parent ($b_{\overline{OP}}$) is drawn. Assume pairs of parents in the parental generation have been selected with their highest phenotypic values. These pairs of parents and their offspring are shown by solid dots. If 'S' be the mean phenotypic value of the selected parents, expressed as a deviation form the population mean, and 'R' be the mean deviation of their offspring from the population mean, then 's' is the selection differential and 'R' is the response. The mean value of the selected parents and of their offspring has the expected position on the regression line (cross mark). The ratio R/S is the slope of the regression time and thus the connection between the response and selection differential is given by

$R = (b_{\overline{OP}})S$

Provided there is no non-genetic cause of resemblance between offspring and parents, $b_{\overline{OP}}$, the regression of offspring on mid-parent is equal to the heritability. Thus

$R = h^2 S$

We know that R, the deviation of the progeny from the population mean is, by definition, the breeding value of the parents. Thus the connection between 'R' and 'S' through h^2 follows directly from the meaning of heritability which says that this heritability is equivalent to the regression of an individual's breeding value on its phenotypic value.

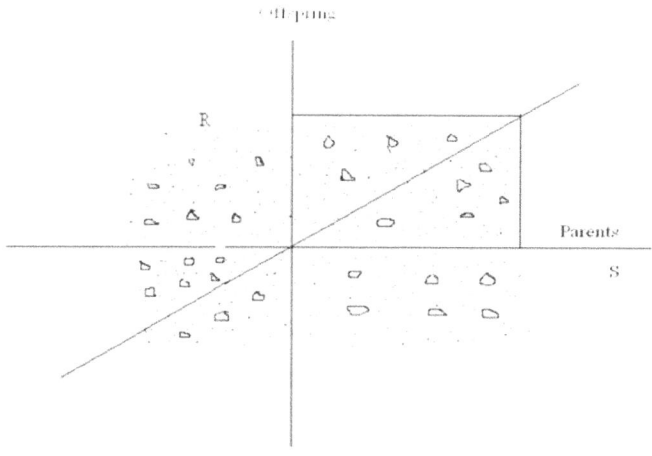

Fig. : Connection between Response (R) and Selection differential (S).

The problem of calculation the mean response to selection is then

To calculate the mean phenotype of the selected groupand

To calculate their mean additive genetic merit or breeding value

Let the phenotype or phenotypic index of an individual be

$I = T + E$

Where,

T = Additive genetic merit of the individual

E = Effects of all other genetic and environmental contributions to the phenotype.

If selection is carried out by ranking the individuals for 'I' and selecting as parents all those whose values exceed 'I', so that a given proportion 'P' of all candidates are selected. If it is assumed that the phenotypes are normally distributed, the selection operation can be represented as follows:

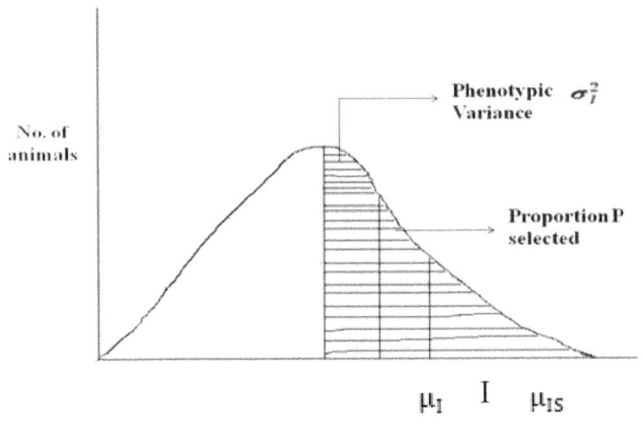

Scale of phenotypic merit

To calculated the mean phenotype of the selected group (μ_{IS}), the above distribution of phenotypes are converted into standard normal variates by subtracting the mean and dividing by the standard deviation to give

$$X = \frac{(I - \mu_I)}{\sigma_I}$$

The variable x is now N (O, 1), i.e. normally distributed with zero mean and variance 1. The distribution of 'x' is as follows:

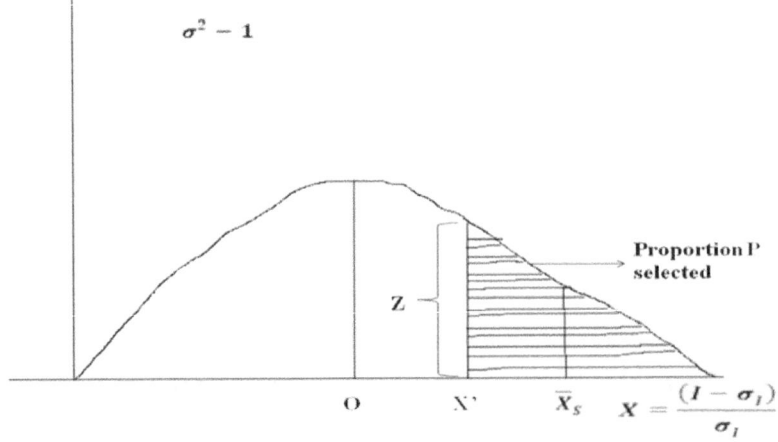

Response to Selection

The selection boundary is

$$X' = \frac{(I' - \mu_I)}{\sigma_I}$$

The mean value of the fraction 'p' or the population falling above this point, X^1, is

$$\bar{X}_S = \frac{1}{P}\frac{1}{\sqrt{2\pi}} \int_{x^1}^{\infty} X^e - \frac{X^2}{2} dx$$

$$= \frac{1}{2}\frac{1}{\sqrt{2\pi}} e$$

$$= \frac{1}{2} f(X^1) = \frac{Z}{P}$$

Where,

Z = the ordinate of the standard normal frequency function, $f(X)$, at the point X^1.

This value for the mean of the selected group in a standard normal distribution is known as the standardized selection differential, $\frac{S}{\sigma_I}$ or intensity of selection, symbolized by 'i' so that $S = {}^i\sigma_1$.

The standardized selection differential $\bar{X}_S = i = \frac{Z}{P}$ can be translated back to the original distribution of phenotypes, μ_{IS} by reversing the previous scaling operation. Thus on phenotypic scale

$$\bar{X}_S = \frac{(\mu_{IS} - \mu_I)}{\sigma_I}$$

$$= \bar{X}_{S\sigma_I} + \mu_I = i_{\sigma_I} + \mu_I = \mu_{IS}$$

Therefore, selecting a proportion 'p' of the population of phenotypes is to produce a group whose mean phenotypic value (μ_{IS}) is higher than the mean of all the individuals (μ_I) in the population. This selection differential, depends on the proportion of the population selected, and on the variability, σ_I, of a normally distributed character.

Now the second stage of the problem is to calculate the mean additive genetic merit or breeding value of the selected group.

Corresponding to the phenotypic distribution, there is an underlying distribution of additive genotypes. The relationship of the two distributions is as under:

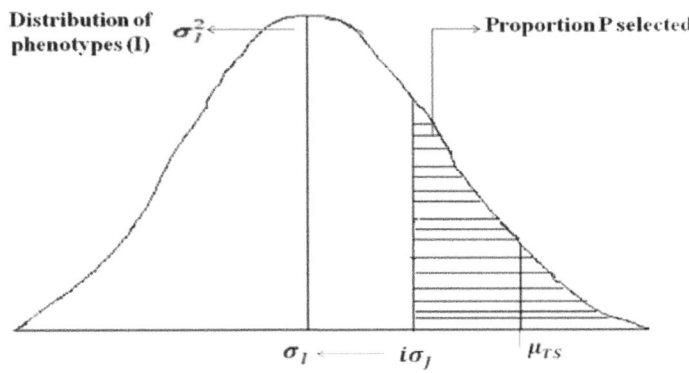

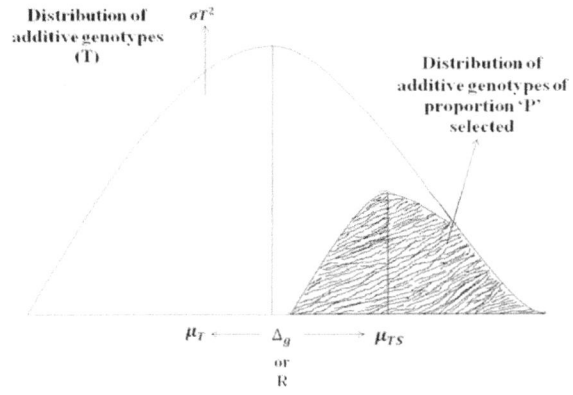

The mean additive genetic value of the selected group is μ_{TS} which is equal to the population mean of additive genotype (μ_T) + the genetic gain (Δ_g) or response to selection (R) resulting from the selection in the phenotypic distribution.

The genetic gain is obtained as

Δ_g = R = (Change in additive genotype per unit change in phenotype) X (amount of change in phenotype)

= (Regression of addive genotype on phenotype) X $^i\sigma_I$ or S

= $b_{TI} (^i\sigma_I)$

= $\dfrac{\sigma_{TI}}{\sigma_I^2} (^i\sigma_I)$

$$= \frac{\sigma_{TI}}{\sigma_I \sigma_T}(^i\sigma_T)$$

$$= r_{TI} \cdot ^i\sigma_T$$

Therefore, the mean additive genetic value (σ_{TS}) of the selected group is

$$\sigma_{TS} = \mu_T + r_{TI} \cdot ^i\sigma_T$$

From the above formula of genetic gain ($Ä_g$ or R =) the three main components of gain identified are:

1. r_{TI} corresponding to the accuracy with which animals are selected. It depends on the heritability of the trait and on the amount of information available for each candidate for selection.
2. 'i' corresponding to the intesnsity of selection, and depends solely on the proportion selected.
3. σ_T corresponding to the additive genetic variance in the population of individuals being selected.

Improvement of Response

The equation R or $\Delta_g = h^2 S = h^2 \sigma_I$ will indicate the ways in which rate of response could be improved. Examining each component of response we see that

1. σ_I, the phenotypic standard deviation, simply specifies the units of measurement and thus has no effect on response.
2. h^2, the heritability, can be increase by
   a. Reducing the environmental variation through rearing and management techniques.
   b. Multiple measurements and
   c. Assortative matings and thus can improve the rate of response to certain extent.
3. 'i', intensity of selection, can improve the rate of response, but two factors limit its scope. They are:
   a). The reproductive rate of the organism: This is because the proportion selected for breeding can never be less than

the proportion needed for replacement. More prolific the organism or sex, more intense the selection. e.g. if males have more offspring than females, change more offsprings than females, and selection can be more intense on males than on females. If the selection applied to males and females differs, the value of 'S' or 'i' will be unweighted means for the two sexes, i.e.

$S = \frac{1}{2} (S_m + S_f)$

$i = \frac{1}{2} (i_m + i_f)$

If, for example, each male mates with 10 females and females have, on an average, 5 daughters each, then the proportion of females to be selected to allow for replacement of the females, can not be less than 1/5, whereas for males having so sons on an average, only 1/50 males need be selected to replace the males. Corresponding to these proportions selected, 'P' the values of intensity of selection, 'i', as a function of 'p', are obtained as $i_f = 1.40$ for females and $i_m = 2.42$ for males, from Biometrika Tables for statisticions and Biometficions II (1931) and reproduced by Becker (1967). The net intensity of i will be i = $\frac{1}{2} (i_m + i_f) = \frac{1}{2} (2.42 + 1.40) = 1.91$.

b). The population size and inbreeding: inbreeding almost always reduces reproductive fitness and the characters related to it. To keep this inbreeding depression at an acceptable level, the number of parents used must be large enough. However, due to limitations of space and other considerations of experiments, the number of parents to be used is fixed, and the intensity of selection can be increased only by measuring more individuals out of which to select the parents.

Generation Interval or Generation Length

Genetic gain per unit time is more important than gain per generation and therefore, the interval of time between generations becomes important. The generation interval (L) is the interval between the matings made in successive generations. It can be calculated as the average age of the parents at the birth of their selected offspring.

Generation interval is lengthened in some forms of selection to the extent that it outweigths the extra advantages of that selecting criterion e.g. in progeny testing by waiting until more offspring have been reared before the selection is made, the accuracy (r_{TI}) and intensity of selection (i) and thereby response per generation can be increased, but in doing so the generation interval is lengthened and will actually reduce the annual rate of genetic gain or response per unit of time. Thus a balance has to be struck between the conflicting interest of intensity of selection (i) and generation interval (L).

If the generation length does not vary systematically among the animals being selected, then gain per year is

$$\Delta g / year = r_{TI} \; ^i\sigma_T / L$$

Where, L is the generation length in years.

When fewer male parents are used than female, the generation intervals of males and females, their intensities and accuracies of selection must be distinguished. The males and females may be further subdivided into sires of males, sires of females, dams of males and dams of females, and each of the four groups may have a different rate of genetic gain and generation interval. If the four genetic gains and generation intervals are $\Delta_1, \Delta_2, \Delta_3, \Delta_4$, and L_1, L_2, L_3, L_4, then not gain per year in the population is

Measurement of Response

When selection has been made for one or more generations, the measurement of the response actually obtained introduces several problems. They are:

1. Generation Means Variable

Because of the variation between generation means, response can not be measured with accuracy until several generation of selection have been made. Assuming that the true response is constant over the period, the best measure of the average response per generation is than obtained form regression of generation means on generation number. The causes of variation of the generation means are: random genetic drift, sampling errors in estimating the generation means, differences in the selection differential and sampling errors can be reduced only by increasing the numbers selected and measured. Environmental differences between generations caused by climate, nutrition, general management etc. can be eliminated by keeping an unselected control population. Assuming environmental differences affecting the selected and control populations alike, the difference between control and selected population estimates the genetic improvement made by selection.

2. Weighting of Selection Differentials

From the genetic point of view the relationship between selection differential and response, and not the response alone, is of interest to use. Further we have to distinguish between the effective selection differential because some parents contribute more offspring than others due to differences in fertility. The selection differential that is relevant to the response observed in the mean of the offspring generations is the effective selection differential which is the weighted mean deviation of the parents. The weight given to each parent, or pair of parets, being their proportionate contribution to the individuals measured in the next generation. The expected selection differential, as defined earlier, is the simple mean phenotypic deviation.

In case differences of fertility (taken in account in effective selection differential) are related to the parent's phenotypic values for the character under selection e.g. more extreme phenotypes are less fertile, then natural selection (differences in fertility) will either help or hinder the artificial selection. By weighting the selection differential the joint effects of natural and artificial selection are measured. Thus, a comparison of effective with the expected selection differential may indicate whether natural selection is operative.

3. Realized Heritability

To show how the response is related to the selection differential, the response is expressed as a proportion of the selection differential i.e. the ratio R/S in the following manner. The generation means are plotted against the cumulated weighted selection differentials. The average value of the ratio R/S is then given by the slope of regression of generation means over selection differentials weighted and summed over successive generations.

Rearrangement of equation $R = h^2 S$, gives.

The heritability estimated in this way from response to selection is called the realized heritability because response may include the effects of systematic changes due to environmental trends or inbreeding depression unless they are removed by comparison with a control line. Selection in two directions may yield very different realized heritabilityies. Although each is a valid description of the response they can not be valid estimates or the heritability in the base population.

Asymmetry of response

Asymmetrical responses have been observed in most experiments and in many two-way selection experiments. The prediction of mean of the responses in the two directions is made from the heritability estimated in the base population. The important practical consequence of the asymmetry is that if there is asymmetry the response in one direction will fall short of expectation. The main causes of asymmetrical responses are:

1. Random drift

When the realized heritabilityies in the two directions are significantly different, the asymmetry is real i.e. not due to random drift. However, if there is only one selection line in each direction, asymmetry can easily result from random drift and it is not easy to prove the reality of the asymmetry. Asymmetry because of random drift can not be predicted as it is expected by chance.

2. Selectional differential

Differences of the selection differential between the upward and down ward selected lines may be due to:

- Natural selection which may aid artificial in one direction or hinder it in the other.
- Change of fertility which will result into higher selection intensity in one direction than in the other
- Change of variance due to change of mean. Increase in variance will cause increase in selection differential and vice versa. This is a scale effect.

While response per generation and the agreement between observed and predicted responses are influenced by differences of the selection differential, the realized heritability is only little affected.

3. Inbreeding depression

If the character selected is subject to inbreeding depression and the selection is made with a small population, there will be decline in the mean through inbreeding resulting into reduced rate of response in the upward direction and increase in the downward direction, giving rise to asymmetry. The amount of asymmetry due to inbreeding can be known by keeping unselected control population with same inbreeding depression. Asymmetry can be predicted with prior knowledge of the rate of inbreeding depression.

4. Maternal effects

Asymmetry of response is shown with traits influenced by maternal effect e.g. selection for weaning weight in two directions in mice. The weaning weight which is maternally determined, hardly increase in the large line while decrease very much in the small lines.

5. Genetic asymmetry

Additive genetic variance and heritability depend on the gene frequencies. Additive genes contribute maximum to the heritability when gene frequency is 0.5 and recessive genes contribute maximum at a frequency of 0.75. These frequencies are called 'symmetrical' gene frequencies. In the initial population, if all the genes affecting the character under selection are at symmetrical frequencies, the realized heritabilities gradually diminish as the gene frequencies get changed through selection, but the diminution will be more or less equal in lines selected in opposite direction and there will be no asymmetry of response. However, if the initial population starts with gene frequencies above or below the symmetrical frequencies, in one selected line the gene frequencies will move away from symmetrical values diminishing heritability while in the line selected in the opposite direction, the heritability will increase as the gene frequency will move towards the symmetrical values. As the gene frequencies become different in up and down lines, asymmetry of response will develop. As this asymmetry is due to acceleration of response in one line and deceleration in the other, it will be associated with non-linear responses. If theoretical limits to selection in two directions are when all alleles influencing the trait under selection, are fixed and if the initial population is not midway between the two limits in phenotypic value, asymmetry of response will result because in one direction the response has to further go than in the other. However, if selection favours heterozygotes, the limit in one direction is equilibrium gene frequency and not fixation. In this case asymmetry will result if the initial population does not have gene frequencies where the additive variance is maximum.

6. Genes with large effects

Asymmetry of response appearing immediately in the first generation of selection can be due to genes with large effects because a large change of gene frequency in produced which is equivalent to many generations of selection on genes with small effects. Asymmetry of this sort is non-linear and thus predictable and occurs when gene frequencies in the initial population are not symmetrical.

7. Scalar asymmetry

The genetic and environmental variation may be skewed to different degrees or in opposite directions in the initial population. Thus at one end of the distribution a larger portion of the total variation will be genetic variation than at the other. This will result into a non-linear offspring parent regression in the base population and asymmetrical response in the first generation of selection. The difference in skewness may be due to genotype-environment interaction (i.e. individuals experiencing good environment exhibit less genetic variation than those in poor environment) or a scale effect (i.e. phenomena viz. a change of variance) following a change of the population mean, are called scale effects. The individuals with high values will exhibit a lower heritability than those with low value, the upword heritability would be grater than the downword.

Chapter - 14

Basis of Selection

Individuality tells us what an animal seems to be; his pedigree tells us what he ought to be, but the performance of his progeny tells us what he is

Effectiveness of selection depends on ability to recognize those animals, which possess superior inheritance. Those superior animals must be mated together for the production of offspring. The aids available to estimate the breeding value of an animal is through the phenotype of an animal or its relatives.

Various basis of selections are:

1. Individual selection or mass selection
2. Pedigree selection
3. Progeny testing and
4. Family selection and sib selection.

Individual Selection

A selection which is based on individual's phenotypic value is called individual phenotypic selection. Individuals are selected solely in accordance with their own phenotypic values. This method is usually the simplest to operate and in many circumstances it gives most rapid response. Mass selection is a term often used for individual selection, especially when the selected individuals are put together in mass for mating. The term 'individual selection' is used more specifically when the mating are controlled for recorded.

This type of selection gives good results when heritability is high, but the effect decreases with falling heritability. It is most commonly sed as basis for selective improvement in livestock. In this method of selection the population is truncated at a particular level of productioj, it refers to a breeding method in which individuals in whom quantitative expression of a phenotype is above or below a certain value are selected as parents for the nex generation.

Therefore it is sometimes called "truncated selection" also. Individual selection is the simplest method to follow for characters which are measured in both sexes. Individual selection, however, also has the following shortcomings :

1. Several importanat traits including milk production in dairy cattle, maternal ability in ewes and sows and egg production in poultray are expressed only by females.
2. Incases where heritability is low, for example milk production, and egg production, the individual merit is a poor indicator of breeding value due to poor accuracy.
3. Most of the carcass demand sacrifice of individuals, hence individual can not be selected for future breeding on the basis of individual performance.
4. The easy and early appraisal of appearance of type, often tempt the breeder to over emphasize the evaluation of selection.

In spite of these shortcoming individual merit is being considered as most important and accurate method of selection.

Pedigree Selection

A pedigree is a record of an individual's ancestors related to it through its parents. Knowledge of the productivity of the ancestors is necessary if pedigree is said to be useful. Such pedigrees are known as performance pedigrees. Ancestors more closely related to the individual should receive most emphasis in pedigree appraisal. The basis of pedigree selection is the fact that an individual gets half of its inheritance from each of the parents and it is usual to expect offspring of outstanding parents to be of higher genetic value than the average of the individual in the herd. Pedigree should be used only as additional information to individual selection.

Indications

Pedigree selection is helpful
- when the trait is sex limited, *e.g.*, milk production, egg production *etc.*,

- when production performances of the individuals are not available,
- for making preliminary selection of sires in progeny testing,
- when the characters are expressed late in life,
- for traits with low heritability pedigree information can be combined with individual's record and
- pedigrees do have the advantage that they are cheap to use.

Limitations

- when the phenotypic value of an individual is known not much is gained by the use of pedigree,
- the genetic make up of the parents can not be known definitely since the phenotype is not indicative of the genotype,
- the pedigree records are made in different environment and hence the accuracy of the ancestry may not be reliable and
- unwanted favouritism towards the progeny of the favoured individual.

Family Selection / SIB Selection

Family, in animal breeding, includes full-sib and half-sib families. In a random mating population, half-sibs have a relationship coefficient of 0.25 and full-sibs have a relationship coefficient of 0.5. Such family members are collaterally related not directly related. They are neither ancestors nor descendants. Because of their common ancestry, they would have some genes in common and thereby some performance in common.

If the records of the individual are included in the family average and used as a criterion for selection, it is known as family selection. If the individuals' records are not included in arriving at the average, then it is known as sib selection. When selection is carried out for market weight in swine, the market weights of all males and females in the family are considered in the calculation of family average (family selection). But when selection is carried out for fertility traits and milk yield, the performance of males can not be included but they are selected on the basis of sibs' average (sib selection).

The families are ranked and based on this, the entire family is selected or rejected. Family/sib selection is used more frequently in swine and poultry where the number of progenies produced by females is high. The family selection does not increase generation interval. The information from family/sib is combined with individual information in the form of index and selection is based on the index.

Indications

- for sex-limited traits,
- for carcass traits and
- for traits of low heritability.

Limitations

- if selection intensity is more, then there may be an increase in inbreeding and
- increase in cost and space in raising larger population.

Precautions

- number of progeny in each family should be large and
- there should not be common environment between sibs.

Within Family Selection

This is the reverse family selection, the family means being given zero weight. The chief condition under which this method has an advantage over the others is large component of environmental variance common to members of a family. Selection with families would eliminate this large non genetic component from the variation operated on by selection. an important practical advantage of selection within families, especially in laboratory experiments is that it economizes breeding space. However the family selection is costly of space. If single pair matings are to be made, then two members of every family must be selected in order to replace the parents. This means that every family contributes equally to the parents of the next generation.

Thus when selection within families is practiced, the breeding space required to keep the rate of inbreeding below a certain value is only half as great as would be required under individual selection.

Progeny Testing

By progeny testing is meant estimating the individuals transmitting ability (breeding worth) by studying its offspring. The principles of progeny test come from the sampling nature of inheritance. Each offspring of a sire caries a sample half of his genes. If enough samples are taken i.e. when enough daughters are measured, we can confidently expect them to average close to the sires true genetic merit.

The term progeny testing is often used in a broad sense to cover two complementary studies. Namely (i) progeny testing in the strict sense of the term i.e. judging the breeding value of the sire by the average performance of his daughters in relation to some standard say herd average and (ii) the evaluation of the sire's merit i.e. estimation of an index of the breeding worth of the sire or sire index. Sire index is of use in the selection of best sires from among those under test.

Testing Procedure

There are three methods of judging superior sires which in brief are as follows:

The most commonly used in the daughter dam comparison which indicates the differences between production level of a sires offspring and their dams. A plus indicates the breeding worth of a sire above the herd average while a minus indicates below average. This method has the drawback that as the dam's as well as the daughters records are generally made over several years, environmental variations over this span of time get confounded with the effect under test.

Another method of evaluating sires is the daughter herd mate comparison practiced in the united state. Herdmates are all daughters of other sires that complete records in same month or the proceeding or succeeding two months as the daughters of sire being evaluated. Since the comparison is made with records of contemporary herdmates, environmental differences are largely eliminated.

The evaluation of sires can be basis of contemporary comparison, a term similar to daughter herdmate comparison with the additional requirement that a herdmate be of the same age as well as calving within the same herd and season as the daughter of the sires being tested. The main advantage of this method over that used in USA is that the age corrections are not required since the daughters as well as their contemporaries start their first lactation at about the sameage. The use of this procedure, however is possible only when enough number of daughters per sire is available.

Requirements For Progeny Testing

At the Experimental Stage

For securing date for progeny testing and sire evaluation, it is necessary that the breeding experiment satisfies certain conditions. In the first place, the sires must be mated to a random groups of dams so as to provide a representative sample of the females constituting the herd to each sires. Care should be taken to group the families into homogenous group according to the available information such as age, performance of the animals etc. and then allote the females in each homogenous group to different sires. It is most essential that a random sample or all, of each progeny group must be raised and recorded for the trait under consideration. Any deliberate selection of data through premature culling of the progeny would vitiate the conclusions of testing.

At the Analysis Stage

Before subjecting the production data generated from the breeding experiment to statistical analysis either for testing or for sire evaluation it is necessary that these are adjusted for any animal to animal variation in the non-genetic factors such as age, lengthof production record etc. It may be stressed that every effort should be made to control the non-genetic variation by properly designing the experiment.

Drawbacks of progeny testing

- To take the long time that it takes to secure the results of performance of the progeny.
- In small closed herd use of progeny tested sires may actually retard the genetic improvement.

Chapter - 15

Methods of Selection

An animal breeder usually interested in various economic traits and when wishes to improve them in the herd/flock, he has three general methods at his disposal. Tandom method, independent culling level method and index selection method.

Tandom Method

It involves selection for one trait at a time and until it reaches at a satisfactory level, and then for a third trait etc. This method is applicable only when the different traits in question are entirely independent or positively genetically associated to each other. Its effectiveness decreases with the increase in number of traits. This method is least desirable of all the methods.

Independent Culling Level Method

Under this systme a culling level of each trait is determined/ fixed at which animal will be culled. The culling level of each trait is determined based on the (1) heritability of the trait (ii) economic importance of each trait and (iii) percentage of the animals to be culled. Here performance in one trait is considered entirely independently from performance of other traits. This system is effective when two traits are considered and (ii) selection pressure is high i.e. when the percentage of animals retained is small. This method is of special consideration as animals can be selected at different stages/phase of growth and development and even at different times of the year. For example in cases of culves and lambs upto weaning or marketing age one can set a level of gains in body weight. The calves or lambs can be selected at an early age based on gaine made by them over the period fixed.

The major limitation of this method is that an animal which is even out standing in all traits except one in which he fall barely below the culling level would be culled.

Index Selection

When selection is applied to the improvement of the economic value of the animals. It is generally applied to several characters simutaneously and not just to one trait because economic value depends on more than one character. This is usually known as multi trait selection. The method that is expected to give the most rapid improvement of economic value. However, is to apply the selection simultaneously to all the component characters together, appropriate weight being given to each character according to its relative economic importance, its heritability and the genetic and phenotypic correlations between the different traits? The component characters have to be combined together into a score of index, in such a way that selection applied to the index, as if the index were a single character, will yield the most rapid possible improvemrnt of economic value.

- A selection index should have the following characteristics:
- It should be simple the underst and and apply by an average breeder.
- It must be based on accurate and comparable records. Here, it is improtant to note that a selection index made for animals in a herd in one environment will be ineffective and inaccurate for comparing animals of the same level in next generation with different of the same level in next generation with different environment. Further, error made in record keeping will affect the accuracy of the results of the index.
- It should be effective i.e. progress thorough the index method should be rapid which would be possible if traits with high heritability are properly emphasized in selection.

Construction of the Index

The index to be constructed for the improvement of merit is, as before:

$I = b_1P_1 + b_2P_2 + \ldots\ldots\ldots + b_mP_m$

Where P_1 to P_m are phenotypic measurements of m characters on which selection is to be based and b_1 to b_m are the corresponding weighting facoors to be determined. To b's are partial regression coefficients of H on I. information from relatives can be included in the index, so the P_s can be measurements of relatives.

Single trait

First consider selection aimed at improving just one character. The purpose of applying index selection is thaen to use secondary charaters as aid to improvemtn of the one desired charater. The equations with charater I as the character to be improved.

$$b_1P_{11} + b_2P_{12} + \ldots\ldots\ldots + b_mP_{1m} = A_{11}$$
$$b_1P_{21} + b_2P_{22} + \ldots\ldots\ldots + b_mP_{2m} = A_{21}$$
$$b_1P_{m1} + b_2P_{m2} + \ldots\ldots\ldots + b_mP_{mm} = A_{m1}$$

The notation here is abbreviated as P_{11} is the phenotypic variance of charater 1 and P_{12} is the phenotypic covariance of character 1 and 2; A_{11} and A_{12} are similarly the additive genetic variance and covariance. The variance and covariances can be expressed in terms of the heritabilityies and correlations, where the subscripts i and j refer to any two different characters and $ó^2$ is the phenotypic variace.

$$P_{ii} = \sigma_i^2 \; ; \; A_{ii} = h_i^2 \, \sigma_i^2$$
$$P_{ij} = r_P \, ó_i \, ó_j \; ; \; A_{ij} = r_A \, h_i \, h_j \, ó_i \, ó_j$$

When the values of the variances and covariances have been entered the soulution of equations provides the values of the weighting factors, b to be used in the index.

Multiple traits

Under this consider simultaneous selection for several characters. The objective is to improve the aggregate breeding value or net werit, which is a particular combination of all the characters to be improved. Merit is now defined as

$$H = a_1A_1 + a_2A_2 + \ldots\ldots\ldots + a_nA_n$$

The relative improtance attached by the breeder to each character. The weihting factors can be economic values; that is to say each a is the value in money units of 1 unit of character.

Economic value

The economic value is the profit made form the sale of the individual. This is then the phenotypic value of merit, which is the character to be improved, and the index is constructed for the improvement of this single character. The economic values of individuals can not be known at the time they are being considered for selection and therefore can not be included as a character in the index.

Accuracy

The correlation r_{IA} between index values and breeding values, which is maximized in the construction of an index, is known as the accuracy of the index. The accuracy of individual selection is therefore the correlation of phenotypic values with breeding values, which is h the square root of the heritability. The accuracy r_{IA} of an index is calculated as follows:

$$\sigma_I^2 = b_1 A_{11} + b_2 A_{21} + b_3 A_{31}\ldots\ldots\ldots\ldots$$

The construction of the index so that the weighting factors, the b's are partial regression coefficients results in the regression of breeding values on index values being unity. Ic, bAi =1. Another way of saying this is that in its construction, the index is scaled so that 1 unit of the index is equivalent ot unit 1, unit of the index is equivalent of 1 unit of predicted breeding value. Now, cov $A_1/ = b_{A1}$ =1 from which it follwos that cov A_1 = . The correlation is given by

$$r_{IA} = \frac{cov_{IA}}{\sigma_I \sigma_A} = \frac{\sigma_I}{\sigma_A}$$

Here ó$_I$ is obtained from equation and ó$_A$ is the square root of the additive genetic variance in the population.

Response to Selection

The response to selection is the mean breeding value of the selected parents, which is predicted form the regression of breeding value on index ualues as

$R = b_{AI} S$,

Where S is the selection differential of index values putting b_{AI} = $r_{AI} \sigma_A / \sigma_I$ and $S = i\sigma_I$ gives the predicted response as $R = ir_{IA} \sigma_A$

General Considerations and Limitations

- The index method is times efficient than independent culling level mentod (where n is the number of traits under consideration in selection programme).
- The progress made for trait by use of the index would be time that made by selection for that trait alone. If suppose, selection were made for 4 traits in an index, the progress for each would be or or 50% as great for each of the traits as compared with progress made when selection is applied only to that one trait, when ofcourse the traits are given equal emphas is.

Chapter - 16

Open Nucleus Breeding System

Open nucleus breeding system is considered to be a best approach to bring genetic improvement in livestock. This system was first introduced New Zealand for faster genetic improvement of sheep. The milk recording under field condition is quite expensive. Moreover, farmers are not under any obligation to retain the animal till the lactation is completed. The generation interval is also very large as it takes about 6-7 year by the time complete record on progeny testing programmes in developing countries is available, so that the effective genetic improvement is small. Use of best proven bull on the elite cow, which are best 5-10% of the cow for genetic improvement of the population. The net genetic gain in the herd considering the actual genetic improvement from various parents off spring path is around 0.7% per annum. Considering these situation in develop countries it has been proposed that 'nucleus' herd be created where males from the best cows are obtained. The nucleus herd would be open in the sense that the lowest yielding cows are culled every year and are replaced by procurement of that many high yielding cow from farmers herd. The nucleus herds are utilized entirely for production of males for breeding purposes in the population.

A nucleus breeding system consists of a population dividing into two tiers:

The nucleus: Composed of genetically elite individuals

The base: forms majority of population.

In this system the selection in practiced in a specialized herd and the females and males are selected on the basis of family information not on that of their progenies. In female selection, information of candidate herself can also be added.The nucleus is of two types:

Open nucleus: The nucleus is said to be open when there is a two-way flow of genes. The replacement may be selected from both nucleus and base borne animals i.e. the transfer of females in both the direction at regulated interval.

Closed nucleus: the nucleus is referred as closed if new genes are not introduced in the nucleus either through male path or through female path. The replacement stocks for the nucleus population are breed entirely within the nucleus and gene can only move in one direction from nucleus to base.

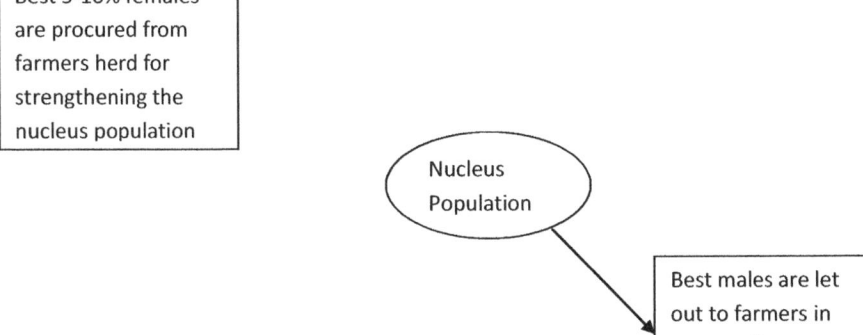

Diagrammatic Representation of Open Nucleus Breeding System

However, to make the people understand the concept and to involve them in breeding programmes at grass root level, the existing farms of State Governments/Central Government should take lead in this direction. These farms have excellent infrastructural facilities and trained manpower to operate the nucleus schemes involving the commercial breeders and progressive farmers.

The MOET was introduced to improve genetic gain under conventional progeny testing schemes by increasing the selection intensity, accuracy of selection and by reducing the generation interval by increasing the number of young bulls. The males are evaluated on the basis of their pedigree information, when they are selected an early age at 12 months or so. This is called "Juvenile MOET". When the males are selected on the basis of the expected performance of their half sister and full sister at the age of around 45-51 months. This is called "Adult MOET". This reduces the generation interval and makes it possible to prove young bull at the age of 3-4 year instead of 7 year in conventional breeding plane.

Although accuracy of such test is lower than progeny information but because of reduction in generation interval, the genetic gain through the use of ONBS would be similar to that of progeny testing.

Multiple Ovulation Embryo Transfer And Open Nucleus Breeding System

Multiple ovulation embryo transfer (MOET) is a composite technology which includes super ovulation, fertilization, and embryo recovery, short-term in vitro culture of embryos, embryo freezing and embryo transfer. Benefits from MOET include increasing the number of offspring produced by valuable females, increasing the population base of rare or endangered breeds or species, ex situ preservation of endangered populations, progeny testing of females and increasing rates of genetic improvement in breeding programmes. Genetic improvement of ruminants in developed countries has made much progress in the last 35 or so years through the use of large-scale progeny testing of males. As has been pointed out, the general failure of extensive use of AI in developing countries has implied that progeny testing schemes cannot be operated with much success. In any case the generally small herds/flocks and uncontrolled breeding in communal grazing situations preclude implementation of progeny testing. Smith (1988a) suggested that the Open Nucleus Breeding System (ONBS) may be especially valuable for developing countries where the use of AI has been a failure due to the reasons given above.

The ONBS concept is based on a scheme with a nucleus herd/flock established under controlled conditions to facilitate selection. The nucleus is established from the "best" animals obtained by screening the base population for outstanding females. These are then recorded individually and the best individuals chosen to form the elite herd/flock of the nucleus. If ET is possible, the elite female herd is used through MOET with superior sires to produce embryos which are carried by recipient females from the base population. The resulting offspring are reared and recorded and the males among them are evaluated using, as appropriate, the performance of their sibs and paternal half sibs and their own performance. From these, an elite group of males with high breeding values for the specific trait is selected and used in the base population for genetic improvement through natural service or AI. It should be noted that, while MOET improves the rate of progress substantially, it is possible to operate an ONBS without ET technology, especially in species, such as small ruminants, with high reproductive rates. Such schemes are being tried for sheep in West Asia by FAO (Jasiorowski 1990) and in Africa (Yapi et al 1994). However, availability of AI and ET, in addition to increasing rates of genetic gain, enhances the flexibility of the system. For example, germplasm from other populations can be introduced easily through semen and/or embryos. One of the advantages of a nucleus herd is that it provides opportunity to record information on more traits than is possible in a decentralized progeny testing scheme. The ONBS can be used for the improvement of an indigenous or exotic breed. It can also be used to improve a stabilized crossbred population. The level of the genetic response depends on the size of the scheme (that is, number of participating herds/flocks and total number of animals) and the selection intensity.

An ONBS can initially be developed to form a focus for national sire breeding and selection activities. In time, and with experience, the capacity can be expanded and ET introduced to increase the rate of genetic progress.

At one time it was suggested that application of MOET in nucleus breeding schemes could increase animal genetic gains by 30-80% (Nicholas and Smith 1983). More recently it has been concluded that the earlier figures were over-predictions (Keller et al 1990). The over-predictions arose partly because the assumed average number of progeny (eight) per donor female was unrealistically high and partly because of wrong assumptions made about genetic parameters (Keller et al 1990). The realistic average number of live progeny per donor flushed is in the range of 2-3 in sheep and cattle and 6-8 in goats (Macmillan and Tervit 1990). Considerations of these figures suggest that MOET could increase annual genetic gains by 10-20% in large nucleus breeding schemes. However, costs of operating such schemes in developing countries need to be evaluated before they can be recommended.

Common Feature of MOET Nucleus Schemes

The basis idea of MOET nucleus schemes is to setup an elite herd of males and females and carryout intensive selection and testing within specialized herd (s) selecting males and females at an early age using family information.

Males are judged on the performance of their sisters and half sisters, this is called sibling test, and not on the performance of their daughter, so generation interval is considerable reduced. Although the accuracy of selection under sibling test is generally lower than that achieved in progeny test, the benefit of reducing generation interval outweighs the loss of accuracy.

As selection and sibling test is done with in a herd or in a few specialized herds, the grater degree of control on the determinant of genetic change i.e. intensity of selection, generation interval and rate of inbreeding is possible.

Variants of MOET Nucleus Schemes

The existing MOET nucleus scheme could be characterized by mainly two criteria.

Kind of information used

Whether progeny testing of males required for nucleus replacement is kept or given up.

If selection is done on ancestors performance i.e. dams plus dams ancestors plus dam's sibs and sire's sibs, it is referred as "Juvenile MOET".

If along with ancestor's performance, the performance of sibs in case of male and sib plus own performance in case of females are considered for selection decision the scheme is referred as "Adult MOET".

If progeny testing of male is continued, than it is called mixed MOET or "hybrid MOET".

MOET nucleus scheme, therefore, could be juvenile MOET or juvenile mixed MOET or adult MOET or adult mixed MOET. The other two characteristics of MOET scheme also need to be mentioned. Whether nucleus is open or closed.

Nucleus scheme should be restricted only at one herd or it May be coordinate in a few herds.

An ONBS for dual purpose cattle

Bioclimatic conditions in the tropics impose serious constraints to the animal production systems, not only on the expression of an animal's genetic potential for production but also on other components. These negative effects should be overcome by using animals well adapted to the existing environment, together with a potential for high production (Frisch and Vercoe 1982); better feeding and management are also needed to ensure the best conditions within the socioeconomic constrains. Cattle production systems show great variability worldwide, but variation among breeds can increase that variability (Dickerson 1969). Different breeding systems that involve crossing have been suggested (Koch et al. 1989) but selection within the cross-breeds is necessary: open nucleus breeding units have been proposed for this (Cunninghan 1981; Smith 1988). These considerations were the framework to develop a breeding system named 'Open Nucleus Cross-breeding Scheme' (ONCS), based on the use of between and within breed variation, for both potential of production and adaptation to the tropical bioclimate through cross-breeding and selection under local conditions in a open breeding unit (OBU) (Osorio 1994).

The Open Nucleus Cross-breeding Scheme (ONCS) that has been shown developed has a population of cross-bred selected cows in its OBU that has shown to have enough genetic potential for production and resistance for the needs of regional dual purpose cattle production systems. This has allowed their offspring to perform well under commercial conditions in our preliminary evaluations.

In India the National Dairy Development Board operates an Open Nucleus Breeding System (ONBS) project for Sahiwal X Holstein Friesian crossbreds since 1994. Bulls are evaluated under a nucleus breeding scheme using Multiple Ovulation and Embryo Transfer as an alternative to progeny testing. The Open Nucleus Breeding System offers the advantage of testing bulls under controlled farm conditions. Since under ONBS sires and dams are evaluated on the basis of their sib performance and not on their progency programme, such evaluations accelerate genetic gain as the generation interval goes down.

Optimal structure of open nucleus system

The principle factors that influence the outcome of an open nucleus system are

- Nucleus size in relation to base population
- Migration rate between tiers

Nucleus Size

James (1977) showed that in cattle and sheep population the highest rate of genetic gain in an open nucleus system was achieved when 5-10% of the population was in the nucleus.

Hopkins (1978) argued that optimum nucleus size with respect to genetic gain is a compromise between a small nucleus resulting in high genetic lag and low selection intensity and a large nucleus resulting in a smaller genetic difference between nucleus and base population, but allowing higher selection intensity.

James (1977) predicted the rate of genetic gain with a variety of nucleus size in an ONS in which male migration was restricted, so that all sires used in both the nucleus and base were nucleus breed thus there is no upward male migration. The predicted selection response increase sharply with increasing nucleus size when nucleus size and small, stabilized at optimum nucleus size and decreased slowly as nucleus size was increased beyond the optimum.

James (1977) demonstrated that as the female selection intensity decreased and male selection intensity increased the optimum nucleus size decreased.

Hopkins (1978) investigated the effects of different age structures and selection methods in the nucleus and base populations. Over the range of parameters studied the optimum nucleus size was generally in the range of 5-10% of the total population. The flatness of the response curves around the optima, however, meant that nucleus size could be increased to 15% without seriously affecting genetic gain.

Migration Rate

James (1977) showed that in a breeding system in which all replacement males, for both base and nucleus, were bred in the nucleus, introduction of approximately half of the female nucleus replacement from the base gave maximum predicted genetic gain. This also approximately doubled the effective population size and thus halved the rate of inbreeding.

James (1977) showed that with increased male selection intensity the optimum proportion selection from the base population increased, but decreased with higher female selection intensities.

Hopkins (1978) determined for optimum migration rate in an ONS in which male migration was not restricted to a downward transfer, but in which both base and nucleus male replacement may be bred in the base as well as in the nucleus. He reported that the base bred male made very little genetic contribution to the nucleus and optimum female migration rates were very similar to those found by James (1977).

Shepherd (1991) evaluated that the highest genetic gain was achieved when approximately were bred in the tier below, and half the male replacements were selected from the tier above.

Some Theoretical Consideration of ONBS

Rate of inbreeding

James (1978) presented a formula for effective size. The effective population size in an ONS is a function of the effective population size in the nucleus and the base and the rate of gene flow between them. He also showed that effective population size is approximately twice the size of the nucleus.

Overlapping generations

Hopkins and James (1978) extended the discrete generation model for describing the overlapping generations. To do this the genetic selection differential account for (1) selection both with in and between parental age subclass (2) the proportional contribution of each parental age subclass to selected progeny.

Change in Genetic Variance

Hopkins and James (1978) base on the assumption that genetic variance and heritability were equal in the nucleus and base population and constant throughout the selection period. In any selection programme and specifically in an open nucleus system these assumptions are not valid.

The process of selection influence genetic variance in two ways:

The increase frequency of favorable allies causes a reduction in genetic variances in long term.

The effect of linkage disequilibrium, which is gradually broken up by recombination, results in a substantial reduction in genetic variance.

In addition, in an ONBS the mixing of animals from partially isolated lines may result in an increase in genetic variance. The heritability would be higher in the nucleus population than the base because of greater efforts to reduce environmental variance.

Optimal Migration Method

Spepherd (1991) developed the model to accommodate more than two tiers, their were also a number of intermediate multiple tiers. Shepherd's model showed that additional genetic gain was achieved with the addition of extra tiers to the ONBS.

Group Breeding Schemes

In most studies of ONBS the base population has been treated as a single population, whereas in practice it is usually composed of a number of wholly or partially isolated subpopulations. In a group breeding scheme each base flock may be owned and managed by a different breeder. The de-Bosque-Gonzalez (1989) found no significance difference in the response achieved in open nucleus system when the base population of 4500 ewes was treated as single entity or as nine separate flocks of 500 ewes, but at smaller base population sizes finite populations effects are likely to influence genetic gain.

Theoretical Results

Theoretical studies have shown that in ONS has a 10-15% advantage over an equivalent closed nucleus in terms of genetic gain. The advantage of an open over a closed nucleus system has been shown to be somewhat greater (by 20-30%) in British Condition, probably because of small population size than in Australian.

The rate of inbreeding in an open nucleus is also expected to be approximately half of that in a closed nucleus of the same size.

Cattle in Tanzania, Kasonta and Nitter (1990) predicted that a breeding system with no nucleus would be more profitable than an ONS, despite higher rate of genetic gain in the ONS system. This was due to higher cost associated with running at intensively recorded nucleus population.

The genetic gain per unit time in open nucleus

$$G = \frac{(1 + Y) C_n + X C_b}{(1 + Y) L_n + X L_b}$$ (James, 1977)

Where,

C_n and C_b were the weighted average genetic selection differentials for nucleus and base respectively.

L_n & L_b were the generation interval for nucleus and base, respectively.

X = Fraction of nucleus females born in base

Y = the proportion of base female born of nucleus

Some of the Advantage of ONBS-MOET Schemes

- Under ONBS-MOET, the evaluation and selection of males and females is conducted with in nucleus herd. So there would be grater degree of control on the determinants of genetic change i.e. intensity of selection, generation interval and estimation of change in levels of breeding.
- Recording of performance will be more accurate.
- It will facilitate measurements of influence of other factors of production of economic importance like feed conversion efficiency, reproductive efficiency disease resistance, body conformation, ease of calving and ease of milking. Such information is not easy and would be very expensive to collect in progeny testing scheme.
- Overall cost of breed improvement programmes is reduced. Fewer facilities are needed to be created for storage of frozen semen doses.
- Breed improvement programme is feasible as recording of farmer's cow is not conducted.
- This system can be useful in developing countries where herd or flock size is small and a short of co-operative effort can be made for establishing a sizeable nucleus.
- Instead of the normal 1% annual genetic progress, as much as 2 to 3% annual genetic progress can be expected at least in the begin years.

Demerits

Farmers may not be ready to part with the best germplasm of females they have-

- Since outside animals are introduced in the nucleus population there is a possibility to introducing some disease in the otherwise disease free nucleus population.
- The success rate of farmer co-operative is usually low.

Conclusion

Open Nucleus Breeding strategy has been chosen to play a leading role in the genetic improvement of cattle. It is a scheme where superior animals are multiplied at a central farm, distributed to farmers and the best animals from the farmers are brought back to the central farm for further breeding and the best bulls recruited for semen production in the Bull stud.

Open nucleus breeding system could result a higher rate of genetic gain and a lower rate of inbreeding then closed nucleus system or other alternatives. The advantage of an open nucleus system is apparently reduced when the cost of running a selection nucleus is high relative to selection cost in the base population or when female reproductive rate is high. For large population the optimum nucleus size is 5-10% of the population size, and approximately half of the female nucleus replacement should be selected from the base population. It is also concluded that with the use of MOET in ONBS will increases the genetic gain by reducing the generation interval and increasing the intensity of selection.The use of multiple ovulation and embryo transfer technology may rapidly multiply the elite germplasm. Therefore the nucleus schemes hold the promises to enhance had to sustain the milk productivity of dairy cattle and thereby help to enhance the socio-economic status of its stakeholders.

A Model of ONBS

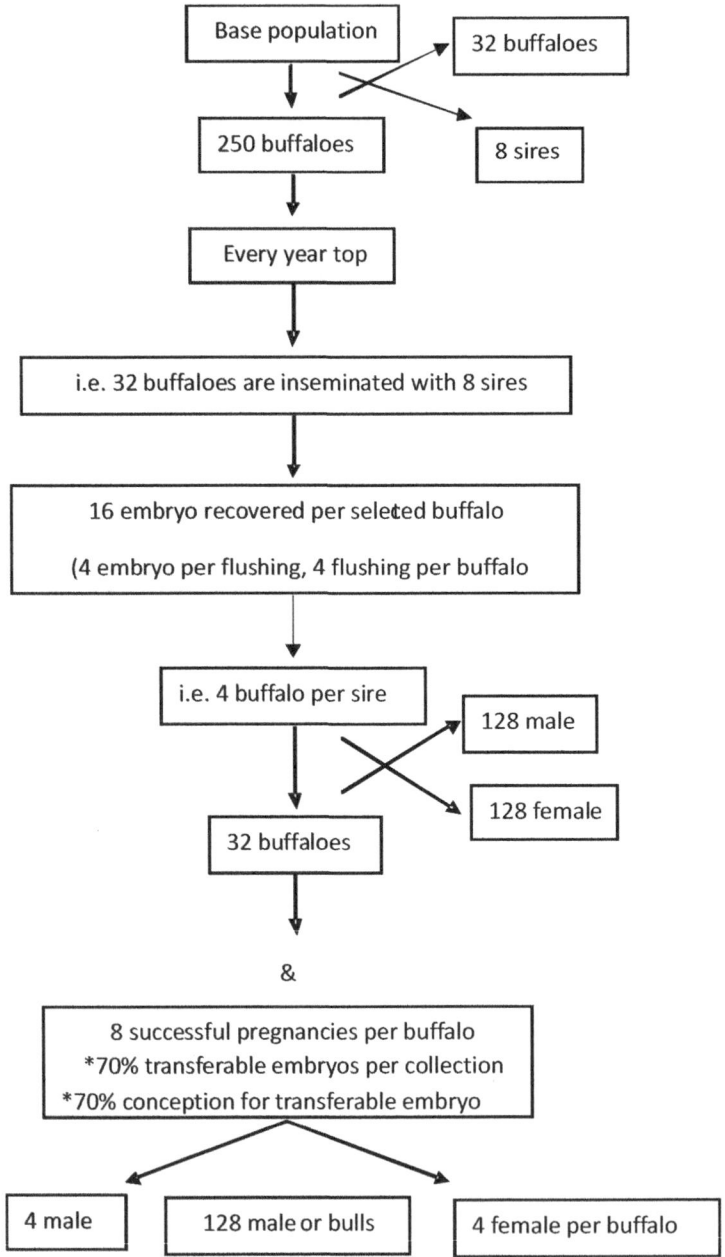

* The males or bulls use for progeny testing and sperm station established under operation flood and also for natural service bull programme

Chapter - 17

Methods of Sire Evaluation

Progeny testing in India was introduced in the late 1950's the first sire index in our country was suggested by Krishnan (1956) which was, however, similar to the herdmate and contemporary comparison methods. Robertson and Rendel (1954) stated that an efficient method of sire evaluation should satisfy two conditions:

- An estimate of breeding value should be unbiased so that the bulls are ranked correctly.
- An estimate should have a minimum variance.

Numerous indices/methods of sire evolution have been used in India. The details of these methods are as under:

Following notations as used by the authors will be used for discussing the indices applied in various studies:-

D	:	Average yield of daughters of the sire
M	:	Average yield of dams mated of the sire
C. D.	:	Average yield of contemporary daughters
C. M.	:	Average yield of contemporary dams
A	:	Herd average
b	:	Regression coefficient of daughters yield on dams yield
h^2	:	Heritability of trait
n	:	Number of daughters of the sire
Q	:	$n/ 1+ (n - 1) 0.25 h^2$
K	:	$(4 - h^2) / h^2$
Y_{ijkl}	:	Observation on lth daughter of the Kth sire in jth

farm in the ith year season of calving

u : Overall mean, YS_1 = Fixed effect of ith year season of calving

F_j : Fixed effect of jth farm

S_{jk} : Random effect of Kth sire in the jth farm

e_{ijk1} : Random error

Simple daughters average index: (I_1)

Proposed by Edward (1932)

$I_1 = D$

It measures the breeding worth of a sire in terms of its progeny performance. For this index it is assumed that-

a – Progeny group size are very large

b – With good distribution over different environmental subclass

c – Equal dams genetic level.

Equivalent Parent or Intermediate Index (I_2)

Proposed by Hansson (1913)

This index is based on the principle that the two parents contribute equally to the genetic make up of the progeny.

This index adjusts the daughters average for varying levels of production of dams. But suffer form the defact that the adjustment tends to over correct, so that the index over estimate the breeding worth if a sire mated to a set of dam's inferior on the average and under estimates the transmitting ability if the dam's happen to be superior on the average to the general level of the herd. Moreover, it is subjected to a relatively high standard error.

Corrected daughter average index: (I_3)Proposed by Krishnan (1956)

$I_3 = D - b(M - A)$

This index adjusts different level of dam's production.

The term b (M –A) allow correlations for non-random allotment of dam's to each sire.

The index proposed by Krishnan (1956) i.e. (I_3) in which the mates level was corrected by the herd average. Which is a constant for a herd? However, if a herd is in existence over a long period of time and there are seasons, period to period differences. Correcting mate's level of production by herd average in not at all correct.

Unbder Indian farm conditions evaluation of bulls will have to be made with -

Information on very few daughters per bull.

Variation in number of daughters per bull.

Records are subjected to serious environmental differences.

Dairy search index: (I_4) Proposed by Sunderson et al. (1965 a)

$$I = A + \left(\frac{n}{n+12}\right)(D - CD) - b(m - CM)$$

The differences due to environmental conditions from period to period were adjusted by substracting daughters and dam's average from their corresponding contemporary average.

This index adjusted different levels of mate's production.

This index also adjusts the number of daughters on which the evolution of sire is based.

The term $\left(\frac{n}{n+12}\right)$ is simply the regression of the performance of present daughters assuming a h^2 of 0.30.

As discussed by Jain and Malhotra (1971) the daughters deviation (D - CD) should be first adjusted for the difference in the dam's production level and then multiplied by the regression factors. Multiplication by $\left(\frac{n}{n+12}\right)$ would estimate the transmitting ability.

For estimating the breeding value the adjusted daughters should be multiplied by because the breeding value is defined as twice the transmitting ability. in view of these arguments this index is also in appropriate for estimating for breeding value.

Contemporary daughter Average index (I_5) Proposed by Jain and Malhotra (1971)

This index takes into accounts the variation in number of daughters in the progeny groups and also corrects for period to period variation in environmental conditions. The correction for period to period variation is made by subtracting form each records the corresponding contemporary average and adding the herd average, whereas the adjustment for the variation in the number of daughters in different progeny groups in ½ h^2 Q is based on the regression of sires breeding worth (G) on D the average of n daughters.

Corrected contemporary average index (I_6)

Proposed by Jain and Malhotra (1971)

This is an extension of index I_5. In addition to adjusting the production records for the differences in progeny numbers and period to period variation also. Adjust for unequal level of dams mated to different sires.

Herd Mate Comparison Method (I_7) Proposed by Henderson et al. (1954)

In this method progeny of sire compared to the other animals in herd of same breed calved in the same years. Herd mate records included the records initiated during a five months interval. The animal which calved two months preceding the months of the daughters calving and the animals which calved during the following two months of daughters calving. For the estimation of sire breeding value the method removes the complications arising form the herd, year, season of calving and feed variation in production.

Predicted Difference Method

Modern trends in sire evaluation in the developing countries are the utilize predicted difference to prove or to estimate their transmitting ability for production (Fairchild, 1979, Young, 1979). To estimate average production of sire's daughter above or below their herd mates (within breed) average was the basic idea of this method. This is comprehended and simplest method for accurate estimates of breeding value of the animals as yield figure are corrected to mature equivalent.

Least squares Technique or ordinary Least Square

The application of least square technique for the estimation and hypothesis testing in data with unequal number of observations has been known since the early 1930's. A compendium of the above method as well as its modification term as ordinary least squares have been given by Harvey in 1966. This method is commonly used for evaluation of breeding bulls in our country with different models using herd, year and year season effects on fixed.

The model given below:
$$Y_{ijk1} = u + Y_{si} + F_j + S_{jk} + e_{ijk1}$$
Where,

Y_{ijk1} = Observation of eth daughter of the Kth sire in jth farm in the ith year season of caving.

u	=	Overall mean
Y_{si}	=	Fixed effects of ith year season of calving.
F_j	=	Fixed effect of jth farm.
S_{jk}	=	Random effect of kth sire in the jth farm.
e_{ijk1}	=	Random error.

Regressed least squares

The regressed least squares have been discussed by Harvey 1979. The regressed least squares can be obtained by-

Regressed least square = least square constant X Weighing factor

Weighing factor = $h^2 / (h^2 + C(4-h)^2)$

C" is the diagonal elements corresponding to ith sire effect of inverted coefficient matrix and h^2 is the heritability coefficient.

Expected breeding value on adjusted data

The data as per least squares analysis adjusted for significant effect of herd year and the breeding value was estimated as per formula of Lush (1935).

$$I = \overline{P} + \frac{2Nh^2}{4 + (n-1)h^2} (\overline{O}N - \overline{P}_5)$$

Where,

is the mean of herd on which sire were tested. is the average of N daughters of the sire.

is the mean of the heard in which progeny of the sire were raised.

N is the number of progeny for the sire tested.

New Method of Indexing Sire Proposed by Narian (1979)

In this index Genetic and Phynotypic relationship of Milk Yield with auxiliary traits e.g. age at first calving and first calving interval were included.

$$I = A + \frac{2nw}{(n+K_y)} \; D - A - \frac{1}{2}h^2(M-A)$$

Where,

$$W = \frac{1 - r_q^2 \dfrac{n+K_{XY}}{n+K_X}}{1 - r_q^2 \dfrac{(n+K_{XY})^2}{(n+K_{XY})(n+K_Y)}}$$

$$K_Y = \frac{(4-h_Y^2)}{h_Y^2}$$

$$K_X = \frac{(4-h_X^2)}{h_X^2}$$

$$K_{XY} = \frac{(Y_{RP} - r_q h_X h_Y)}{r_q h_X h_Y}$$

r_p and r_g are phenotypic and genetic correlations between X and Y traits respectively.

h_Y^2 = heritability of dependent traits (Y)

and

h_X^2 = heritability of auxiliary trait (X)

Best Linear Unbiased Predictor Method (BLUP)

Among the sire evaluation methods, the herd-mate comparison (Henderson et al. 1954) and contemporary comparison (Robertson and Rendel, 1954) are commonly used in 1960's several assumptions made for the herd-mate and contemporary comparison methods become invalid with the introduction of intensive artificial insemination and differential usage of sires. The main limitations of these two old methods of sire evaluation are as follows:

The procedure of taking the deviation of contemporary average from the progeny group average did not adjust the progeny group average effectively for the environmental effects because the herd-mate or contemporary average comprised the confounded effects of herd and sires.

The genetic merit of sires of contemporaries was ignored correction for the genetic merit of sires of contemporaries would help in effective adjustments of estimates of breeding values for genetic trends. Corrections for genetic merit of contemporaties were later included in the cumulative differences (CD) method, suggested by Bar-Anon and Sacks (1974).

The genetic and environmental trends were not accounted for and all proven and unproven sires were assumed to be from a single popn.

The estimates of breeding values were biased by selection because the progeny with more than one lactation were also included in evaluation of breeding values. Such progeny are assumed to be selected since their retention in the herd is generally dependent on their performance upto the proceeding lactation.

The above limitations were overcome in the BLUP procedure of Henderson (1973) presently. The BLUP is the method of choice for evaluation of sires and cows in many countries.

With the recent advances in the application of statistical genetics for analysis of animal breeding data, the least squares, simplified regressed least squares; contemporary comparison methods are absolute except BLUP (Chauhan, 1991). It is also equally important to choose an appropriate model for sire evaluation by BLUP procedure. There is no standard model that could be used in all breeding systems. The importance of different environmental effects in a model could be examined by partitioning the total sum of squares, and by fitting reduced models Henderson (1975), Chauhan and Thompson (1980), Chauhan (1987, 88) have used several criteria for comparison of models. Adequate attention should be given to detect the confounding of effects. The differences due to an effect may be insignificant, but when it is ignored the variation due to such effects may be confounded with sire effects, and thus might bias the estimate of breeding values.

Simultaneous estimation of all environmental and genetic effects is considered most desirable; however, some environmental effects such as age at calving and month of calving are conventionally precorrected using standard correction factors. Chauhan (1988) has recommended that additive correction factors should be preferred for sire evaluation with a linear model, and the correction factors should be derived form a reasonable size of data set.

In view of sire X environmental interactions it is often argued the different sets of bulls would be appropriate for different locations. However, the reports form Europe (Robertson et al., 1960) has showed that sire X herd interactions for milk production traits were small. Chauhan (1983) suggested that sire X environmental interactions were unimportant for milk yield. Therefore, sire X environmental interactions must be tested before selecting the model for sire evaluation in BLUP procedure. If the model is appropriate, the following can be stated for the BLUP procedure:

Predictions of random effects (e.g. sire genetic value) are unbiased, as is the estimation of any estimable function of fixed effects (e.g. age of dam). The expected value of the predictor is the same as that of the predictant.

The variances of errors of prediction are minimized. No other procedure using linear functions of the observations can be found which gives less variance of the difference between predicted and true value (the requirement for a best procedure).

The correlation between predicted and true value is maximum.

If the observations are random variable, have a multivariate normal distribution then the probability of correctly ranking pairs of random variables is maximized.

Prediction is the same as selection index prediction, assuming true known effects, except that the generalized least squares estimates of fixed effects are used to adjust the data.

Limitations of BLUP

It should be pointed out that only linear functions of observations are used and that certain biased procedures may have smaller squared prediction errors.

The correct ration of R and G should be known. However, the incorrect ratio in BLUP procedure is not seriously affected within limits (Slanger and Henderson, 1975).

An equation is needed for every level of every factor. In animal breeding data this may mean thousand of equations and unknown quautitus to be estimated for predicted. Therefore, efficient absorption, itesotion and sortin procedures are valuable toos in solving such age.

Use of BLUP Procedure

Genetic differences between herds –

In this method, the basic unit of comparison of predicting sire effects is to compare the daughters of one sire with contemporary daughters of other sire in the same herd. This procedure appears to be nearly the herd-mate method. The difference, however, can he explained by following illustration (Danel et al., 1976).

Sire	Herd			
	1	2	3	4
A		*	*	
B		*		*
C			*	*
D	*		*	

In herd 1, daughters of sire A can be compared with daughters of sire D, daughter of sire D also project in herd 3, where his daughters can be compared with daughters of sire C, Which indirectly. In the whole body of data every sire is compared with the other sires, either directly or indirectly. Thus, every comparison is made against a known genetic level. However, in herd-mate method the comparison is made against the average genetic level of the herd-mates, which varies from herd to herd, whereas sire comparison in BLUP dominates the bias caused by genetic differences between sires of herdmates.

To use this method, there must be overlapping between herds in bull usage. Furthermore, there must be atleast from different bulls in each herd. Selected bulls (Proven sires) which have daughters distributed in many herds, can act as linkages among young bulls which have daughters in few herds. Definition of the herd effect requires a balance between ideal situation which is to include only daughters of the same age of calving during the same month or season (i.e. dividing the herd into herd, year and season HYs) and the need to have sufficient cows in each Hys to insure at least two sires are represented.

A HYs interval over a greater period of time will in variably lead to the situation where there are many HYs with only one daughter represented. Such daughters will not contribute to the comparisons. On the other hand, a HYs defined over a limited period is a guarantee that the comparisons are possible. The influence of season, in particular, is known to be rather variable from one herd to another and from one year to another. Therefore the definition of the herd-year-season effects resulting from splitting each year into atleast two seasons preferred, even if some daughters are thereby lost and the no. of cows in a HYs is consequently rather low. Including older cows in the data leads to more cows per HYs and better connection between bulls. Adjustments for lactation number are, however, needed.

Using records of progeny of daughters of the sire to be evaluated. The problem with non-random mating can be solved partially by including the effect of a maternal grand sire in the model. When this is done, bulls mated to better cows are less likely to be over evaluated, and vice-versa.

Number of daughters and grouping of sires-bulls to be evaluated can be considered as belonging to different sub-populations. For ex. Bulls indifferent generations can not always be regarded as members of the same populations, because genetic differences between generations. In each sub population of bulls, the breeding value have to eh expressed as difference from their own mean. Otherwise, bulls in a genetically better group will be under-evaluated and vice-versa. In the BLUP method group effect is often include in the model to account for subpopulation mean. A group, therefore, consists of bulls which are genetically more alike than other bulls. There has been much discussion of how the group should be defened since many definitions is rather arbitrary. One way is to group the bulls according to either year of birth of year of testing, or pedigree estimates and so on. According to Schaeffer (1975) there should be 20-40 sires per group.

Estimation of genetic and environmental trends – the BLUP method yields prediction of breeding values for bulls of different generations which are related to the same genetic base. It is, therefore, possible to rank not only the last generation of bull but also to rank, for e.g. the last five generations. In order to obtain this trend, the generation must be presented in the data to be analysed. The bulls which are used during any one year in a modern cattle breeding programme consists fo the latest generation of young bulls and a group of selected bulls which belong to generation 4-5 years older. There must be some overlapping between generations. But if each bull is used over all shorter period, the overlapping will be poor (connection will be poor). Improve connections are achieved when several bull are used simultaneously, with overlapping and over a longer period. The fact that all breeding values are related to the same genetic base makes it possible to calculate the genetic progress made in population. The program from selection of the young bulls for sampling can be estimated as the difference in over age predicted breeding values between different groups of bulls. The genetic trend in the population depends mainly upon how the bulls are used after the progeny test is finished.

The estimated genetic level of for a specific year is simply a function of the weighted average of the predicted breeding values of the sires of all cows freshining for the first time in that year. The difference between averages of different years constitutes a measure of genetic trend.

Genetic trends provides a continuous check on genetic programme in the population as well as check on the genetic level of each sample of young bulls (sets) must be considered as a valuable attribute of the sire comparison method (i.e. BLUP method). The environemtnal trends can also be estimated by BLUP method as illustrated by schoeffer (1974).

One approach is to suppose that the sub-class number have same sort of distribution and are corrected with certain random elements of the model.

The second approach involves a set of linear functions that describe the selection process.

Chapter - 18

Breeding or Mating Systems

The aim of mating systems are either to increase or decrease the homozygosity of the progeny compared with the parents or in some cases, to maintain the degree of homozygosity unchanged. This aim constitutes the basis of the following classification of mating systems:

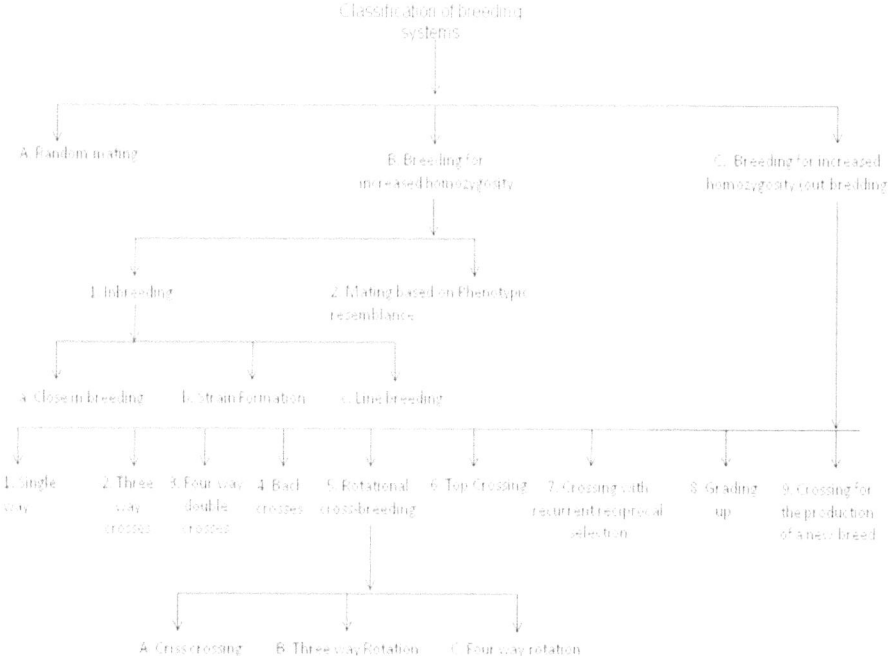

Random Mating

Many organisms in nature mate at random or nearly so. Random mating or panmixia, mean that, in case of bisexual organisms, any one individual of one sex has an equal chance of mating with any individual of the opposite sex in the population. In random mating the probability of the mating of a given female to a particular male is in direct proportion to the different types of males in the population. Random mating with respect to a trait exists when an individual which possesses this trait is no more or no less likely to mate with another possessing the trait than would be expected form the frequency of individuals of the opposite sex possessing that trait in the population.

Breeding For Increased Homozygosity

This very often means mating within a breed. Under this system of mating inbreeding will be dealt in detail while for mating based on phenotypic resemblance it is suffice to mention that under this system no regard is paid to the degree of relationship and the mates are selected only according to the external resemblance of the animals viz. mating large bulls with large cows, small bulls with small cows etc.

Before dealing with inbreeding it would be proper to understand the meaning of 'relationship' in animal breeding.

Relationship

Being related means that two individuals have one or more common ancestors. In farm animals, however, we use the term related in a more restricted sense to mean that the animals mated are more closely related than average animals of their breed, this usually means that there are common ancesoors at least in the first 4 to 6 generations of their pedigrees.

Inbreeding

Inbreeding is a system of mating in which the mates are more closely related than the average relationship between all individuals in the population or breed from which the mates come. The progeny so obtained from this mating is known as inbred. These inbreeds tend to become more homozygous. This increase in homozygosity and the accompanying decrease in heterozygosity is the underlying reason for the genotypic and phenotypic changes associated with inbreeding.

Three main types of inbreeding systems can be distinguished:

Close inbreeding: Examples are:

(1) Mating between sibs

(2) Mating between parents and progeny (incest)

(3) Strain formation

Strains are group of animals within a breed or variety more or less isolated form each other due to geographic conditions and/or developed with some special characteristics in view of different aim of breeding. Inbreeding is considerable milder in strain formation.

Line Breeding

Line breeding is inbreeding within an ancestral line with the object of increasing a particular male or female ancestor's proportion of the genetic constitution in the progeny. The most intensive form of line breeding is back crossing to the same parent for several generations in succession. Usually, a much milder form of line breeding is employed, such as mating a female with a grand sire, or uncle.

Genetic effects of inbreeding

Inbreeding increases the proportion of gene pairs that are homozygous and decreases the proportion that are heterozygous. Increase in homozygosity in inbreds decreases dominance effect in inbreds while heterozygotes may show over dominance i.e. perform better than either homozygote in case over dominance is positive. Basis for indbreeding effects are the dominace and over dominance theories.

Inbreeding itself does not change gene frequencies, but heterozygosity may be lost as inbreeding progresses because in small populations the gene frequency would fluctuate rather extremely and by chance certain genes would be lost as other becomes fixed (homozygosis). As inbreeding progresses without selection, a large population would become a series of sub-populations, a large population would become a series of sub-populations, each likely to be homozygous for a number of different alleles.

Phenotypic effects of inbreeding

Determinatal effects as a consequence of inbreeding are reported mostly on swine and poultry. The traits affected are: (i) Growth rate (ii) Reproductive performance and (iii) Vigor.

Also as a consequence of inbreeding hereditary lethal or other abnormalities appear as these recessive traits get exposed due to increase in homozygosis which were otherwitse covered up in heterozygote combinations.

Use Fullness of Inbreeding

Inbreeding helps in producing seed stock which can be used with predictable results as parents for out breeding or crossbreeding for producing high yielding hybrid animals for commercial production. This technique is exploited more widely in poultry. Other uses of inbreeding are for (i) line breeding (ii) elimination of undesirable recessives (iii) development of families (iv) development of homozygosity and prepotency.

Measurement of Inbreeding & Relationship

Coefficient of Inbreeding

The intensity of inbreeding as measured by the inbreeding coefficient 'F' was proposed by Wright 1921. Coefficient of inbreeding is relative to a particular breed or population at a specified time. Thus the inbreeding coefficient, represents the probable increase in omozygosity which has occurred singe the reference data as a result of the mating of individuals more closely related than the average relationship for the population. Inbreeding coefficient can also be defined as the probability that the two allels at a locus in an individual are identical by descant (two genes are said to be identical by descent if they arise from the direct replication of the same gene from a common ancestor in an earlier generation). Inbreeding coefficient ranges in value from 0 to 1.0. It is also often referred to a percentage ranging form 0 to 100. As the value of 'F' increases, the relative proportion of heterozygous loci declines in a proportionate amount equal to 1 − F.

Calculation of inbreeding coefficient

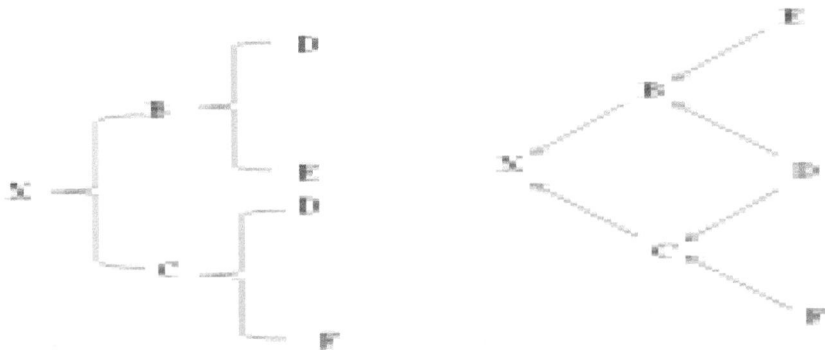

Fig. 1 : Bracket and arrow style pedigrees of an animal (X) produced by mating of a half-brother and half sister.

Consider the pedigree in Fig. 1. We are interested to determine the inbreeding coefficient of individual X. In other words we want to determine the probability that X received genes that were identical by descent from D. transmitted through both B and C or what is the probable proportion of loci in X that is homozygous because X received two replicates of a gene at these loci from D.

The formula for calculating inbreeding coefficients is as follows:

$$F_X = \frac{1}{2}\Sigma\left(\left(\frac{1}{2}\right)^n (1+F_A)\right)$$

Where,

- F_X : Inbreeding coefficient of individual X
- Σ : Greek symbol for summation or adding all paths
- n : Power to which one-half must be raised. Depending upon the number of arrow connecting the sire and dam through the common ancestor.
- F_A : Inbreeding coefficient of the common ancestor.

If the common ancestor is not inbred the formula of inbreeding coefficient becomes:

$$F_X = \frac{1}{2}\Sigma\left(\left(\frac{1}{2}\right)^n\right)$$

In Fig. 1 the sire (B) and dam (C) of individual X has the same sire (D). 'D' is the only common ancestor in the above pedigree because it appears in the pedigree of both the sire (B) and the dam (C) of individual X. The arrow diagram shows only one path way from D to X through the sire (B) and only one through the dam (C). This pathway may be straightened out as below for illustrative purpose:

```
                    1              2
X  ←———— B  ←———— D  ————→ C  ————→ X
```

We now number the arrows running from the sire (B) through the common ancestor (D) to the dam (C). We do not count the arrows running from individual X to the sire and dam. In this case number of arrows connection sire and dam with the common ancestor is two. And this is the 'n' in the formula.

Thus

$$F_X = \frac{1}{2}\left(\frac{1}{2}\right)^2 = (0.25) = 0.125$$

i.e. the inbreeding coefficient of individual 'X' is 0.125 which can also be expressed as 12.5% by multiplying the inbreeding coefficient by 100.

Example: Calculate the inbreeding coefficient from the following pedigree-and -arrowdiagram of a full sib mating:

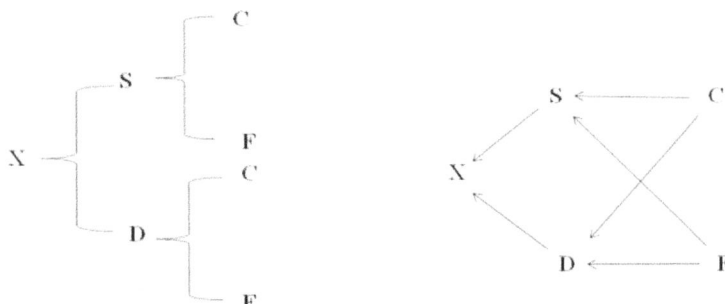

Pedigree diagram   Arrow diagram

The solution

The pathway is for

(C) as common ancestor:

$$X \leftarrow S \xleftarrow{1} C \xrightarrow{2} D \rightarrow X \quad \left(\frac{1}{2}\right)^2 = 0.25$$

(F) as common ancestor

$$X \leftarrow S \xleftarrow{1} F \xrightarrow{2} D \rightarrow X = \left(\frac{1}{2}\right)^2 = 0.25$$

Total = 0.50

Example: Calculate the inbreeding coefficient from the following pedigree and arrow diagram of a father-daughter mating.

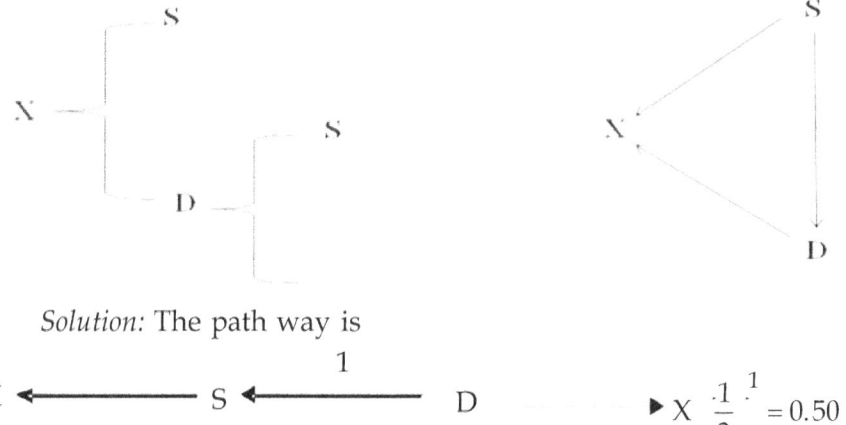

Solution: The path way is

$$X \longleftarrow S \xleftarrow{1} D \qquad \blacktriangleright X \;\frac{.1}{.2}\cdot\frac{1}{\cdot} = 0.50$$

Thus $F_X = \dfrac{1}{2} = (0.50) = 0.25 = 25\%$

Note: The inbreeding coefficients for dam X son matings are calculated in a similar manner. Except the arrow diagrams run from the dam as the common ancestor

Example: Calculate the inbreeding coefficient from the following pedigree and arrow diagram of a sire x daughter mating with the sire inbred.

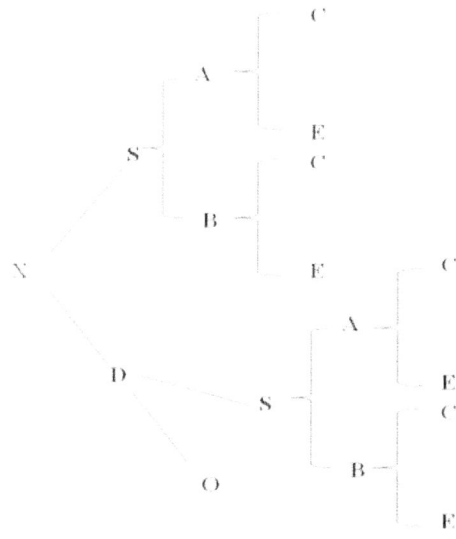

Solution:

The common ancestor in the above pedigree is 'S', which is the sire of individuals X and D. But why only 'S' and not A, B, C and E are taken as common ancestors? The answer is we take care of A, B, C and E by first calculating the inbreeding coefficient of individual 'S' or the sire in this case, as usual. After this is done, calculate the path taking 'S' as common ancestor. Then use the formula of inbreeding coefficient as

$$F_X = \frac{1}{2}\Sigma\left(\left(\frac{1}{2}\right)^n (1+F_A)\right)$$

Inbreeding coefficient of individual 'S'
The pathways with C and E as common ancestor:

$$S \leftarrow A \leftarrow\overset{1}{\quad} C \overset{2}{\quad}\rightarrow B \rightarrow \quad S = \left(\frac{1}{2}\right)^2 = 0.25$$

$$S \leftarrow A \leftarrow\overset{1}{\quad} E \overset{2}{\quad}\rightarrow B \rightarrow \quad S = \left(\frac{1}{2}\right)^2 = 0.25$$

$$\text{Total} = 0.50$$

Thus
Inbreeding coefficient of individual X
The pathway

$$X \leftarrow S \overset{1}{\leftarrow} D \rightarrow X \quad \frac{.1}{.2}\cdot\frac{1}{} = 0.50$$

Since S, which is the common ancestor, is inbred, we must use the complete formula as given above:
Thus

$$F_X = \frac{1}{2}(0.50)(1+0.25)$$

$$= \frac{1}{2}(0.50(1.25))$$

$$= \frac{1}{2}(0.625)$$

$$= 0.3125 \text{ or } 31.25\%$$

Example: Calculate the inbreeding coefficient from the following pedigree and arrow digram of a sire and grand-daughter mating.

      Pedigree diagram   Arrow diagram

See that animals B. E and F are the common ancestors in the above pedigree. Also 'B' is himself inbred, since he is the result of a half-brother-sister mating. Had the inheritance of E and F passed to X only through 'B', there would have been no need of considering E and F as common ancestors because their contribution would have taken care of by calculating the inbreeding coefficient of 'B'. But here E and F are also taken as common ancestors because their inheritance also passes through individual other than B.

Inbreeding coefficient of individual B

    The pathways with F as common ancestor

$$3 \leftarrow D \xleftarrow{1} F \xrightarrow{2} E \xrightarrow{} D = \left|\frac{1}{2}\right|^2 = 0.25$$

Thus $\quad F_B = \frac{1}{2}(0.25) = 0.125 = 12.5\%$

Inbreeding coefficient of individual X

The pathways with B, E and F as common ancestor

(B) as common ancestor

$$X \longleftarrow B \xleftarrow{1} I \xrightarrow{2} C \longrightarrow X = \left(\frac{1}{2}\right)^2 = 0.25(1.125) = 0.28125$$

$$X \longleftarrow B \xleftarrow{1} J \xrightarrow{2} C \longrightarrow X = \left(\frac{1}{2}\right)^2 = 0.25(1.125) = 0.28125$$

(E) as common ancestor

$$X \longleftarrow B \xleftarrow{1} E \xleftarrow{2} J \xrightarrow{3} C \longrightarrow X = \frac{\cdot 1\ \cdot^{3}}{\cdot 2\ \cdot} = 0.125$$

(F) as common ancestor

$$X \longleftarrow B \xleftarrow{1} D \xleftarrow{2} F \xrightarrow{3} E \xrightarrow{4} J \xrightarrow{5} C \longrightarrow X = \frac{\cdot 1\ \cdot^{5}}{\cdot 2\ \cdot} = 0.03125$$

$$\text{Total} = 0.71875$$

Thus $F_X = \frac{1}{2}(0.71875) = 0.359375 = 35.93\%$

Note: In the pathways shown above B is listed twice because the same gene or genes from B might have been transmitted to X directly and also indirectly by either of two paths through J and through I.

Problems:

Calculate inbreeding coefficient of X from the following pedigree and arrow diagram:

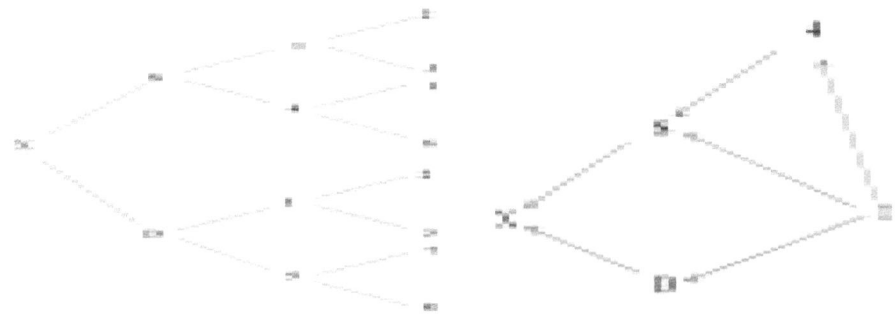

Calculate inbreeding coefficient of 'H' from the following arrow diagram.

Calculate inbreeding coefficient of 'X' from the following pedigree and arrow diagram

Coefficient of Relationship

Degree of relationship as measured by coefficient of relationship was developed by Wright (1921). Relationship coefficient R measures the probable proportion of genes that are the same for two individuals due to their common ancestry, over and above that in the base populations. It should be kept in mind that the relationship between two individuals is the extra similarity in the genes they possess due to their common ancestry. Many of their genes will already be alike because of the high frequency of these genes in the population (breed).

Relationships are of Two Different Kinds

Collateral relationship:

Animals that are related because they are descendants of some of the same animals e.g. half-sub (one parent common). Fully-sib (two parents common), first cousin (one grand parent common), single first cousin (two grand parents common), double first cousin (four grand parents common) etc.

Direct relationship

Relationship between individuals when one is a descendant of the other, e.g. parent of offspring. Grand parent and grand child etc.

Relationship coefficient between collateral relatives formula

$$R_{XY} = \frac{\Sigma\left(\left(\frac{1}{2}\right)^n (1+F_A)\right)}{\sqrt{(1+F_X)(1+F_Y)}}$$

Where,

R_{XY}	=	Relationship coefficient between animals X and Y
"	=	Greek symbol for summation
n	=	Number of arrows connecting individual X with Y through the common ancestor for each path
F_X	=	Inbreeding coefficient of animal Y
F_A	=	Inbreeding coefficient of the common ancestor.

Because inbreeding increases homozygosity, any inbred animal will transmit similar genes to each of his offspring more frequently than will a non- inbred individual. If an inbred animal is the common ancestor of two related individuals, they will therefore have more genes in common and thus be more highly related than if the common ancestor had not been inbred. To take care of this, the contribution of each inbred common ancestor must be multiplied by $1 + F_A$.

Inbreeding also makes a population more variable by producing separate inbred strains. Inbred descendants of any animal will be homozygous in a greater percentage of their gene pairs than if they were not inbred, but they may be homozygous for different alleles of the same gene pair and thus less related than if they were not inbred. The denominator for the relationship formula takes this into account, making the complete formula mentioned above.

If individuals X and Y and their common ancestor not inbred, the formula becomes:

$$R_{XY} = \Sigma\left(\left(\frac{1}{2}\right)^n\right)$$

Example: Calculate relationship coefficient between half-sibs form the following pedigree and arrow diagram.

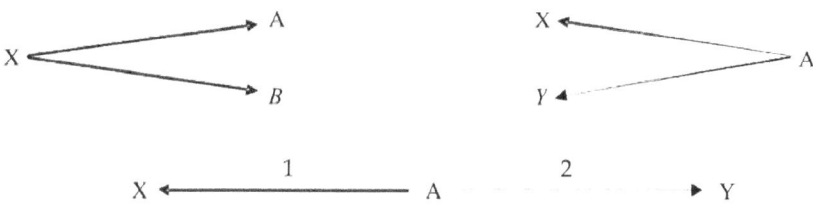

$$R_{XY} = \left(\frac{1}{2}\right)^2 = 0.25 = 25\%$$

Example: Calculate relationship coefficient between full sibs form the following pedigree and arrow diagram.

Pedigree diagram   Arrow diagram

$$X \xleftarrow{\ 1\ } A \xrightarrow{\ 2\ } Y = \left(\frac{.1}{.2}\right)^2 = 0.25$$

$$X \xleftarrow{\ 1\ } B \xrightarrow{\ 2\ } Y = \left(\frac{.1}{.2}\right)^2 = 0.25$$

Total   = 0.50

Pathways

$$R_{XY} = \left(\frac{1}{2}\right)^2 + \left(\frac{1}{2}\right)^2 = 0.25 + 0.25 = 0.50 = 50\%$$

Example: Calculate relationship coefficient between first cousins form the following pedigree and arrow diagram:

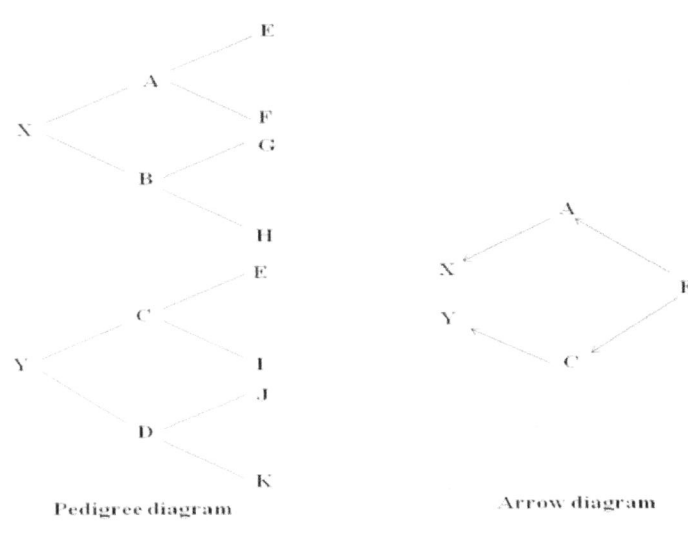

Pedigree diagram   Arrow diagram

$$X \xleftarrow{1} A \xrightarrow{2} E \xrightarrow{3} C \xrightarrow{4} Y = \frac{1}{2} \cdot ^4$$

$$R_{XY} = \left(\frac{1}{2}\right)^4 = 0.0625 = 6.25\%$$

Example: Calculate relationship coefficient between S and D from the following pedigree and arrow diagram

Breeding or Mating Systems

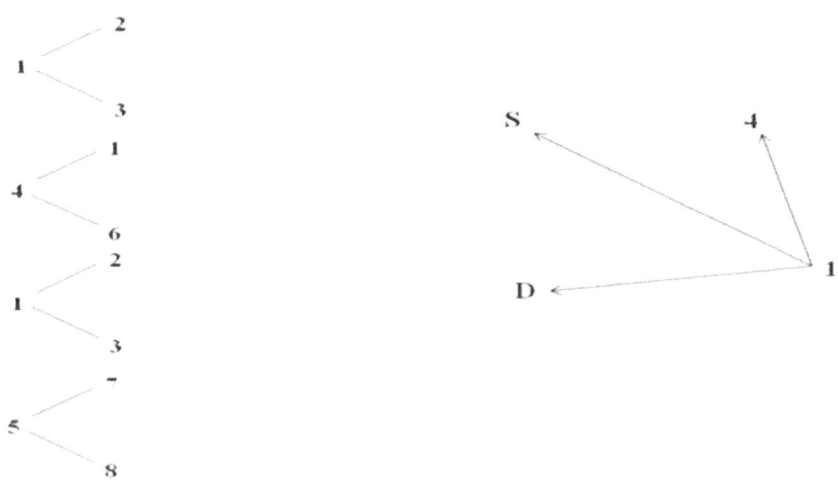

Pedigree diagram   **Arrow diagram**

$$S \xleftarrow{1} 4 \xleftarrow{2} 1 \xrightarrow{3} D = \frac{.1}{.2}^3 = 0.125$$

$$S \xleftarrow{1} 1 \xrightarrow{2} D = \frac{.1}{.2}^2 = 0.0250$$

Total = 0.3750

Pathways
Inbreeding coefficient of S

$$\text{Pathway} = S \xleftarrow{} 4 \xleftarrow{1} 1 \xrightarrow{} S = \frac{.1}{.2}^1 = 0.50$$

$$F_S = \frac{1}{2} = (0.50) = 0.25$$

Relationship coefficient between 'S' and 'D'

$$R_{SD} = \frac{0.3750(1+0)}{\sqrt{((1+0.25 x 1+0)}} = \frac{0.3750}{\sqrt{1.25}} = \frac{0.3750}{1.1180} = 0.3354 = 33.5\%$$

Relationship coefficient between direct relatives formula

$$R_{XA} = \Sigma\left(\frac{1}{2}\right)^n \sqrt{\frac{1+F_A}{1+F_X}}$$

Where,

R_{XA}	=	Relationship coefficient between descendant (X) and ancestor (A)
"	=	Greek symbol for summation
n	=	Number of arrows connecting individual A with X for each path
F_A	=	Inbreeding coefficient of the ancestor
F_X	=	Inbreeding coefficient of the descendant when neither of the individuals involved is inbred, the formula becomes

$$R_{XA} = \Sigma\left(\frac{1}{2}\right)^n$$

Example: Calculate relationship coefficient between sire (S) and his daughter (D) form the following pedigree and arrow diagram:

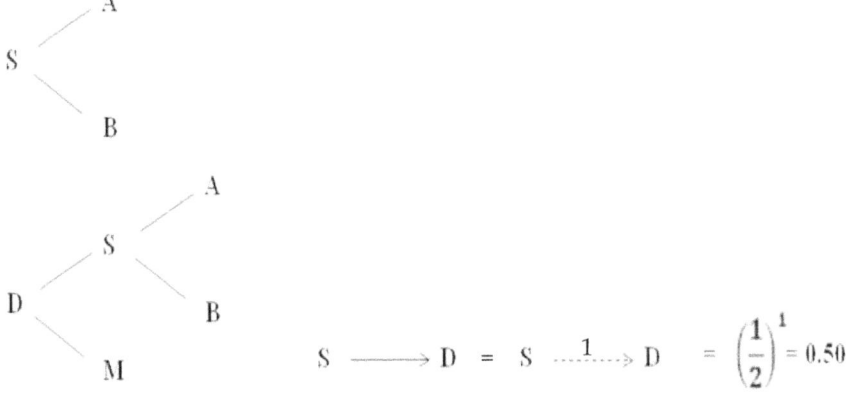

Pedigree diagram   Arrow diagram   Pathway $R_{DS} = 0.50$

Example: Calculate relationship coefficient between X and 1 from the following pedigree and arrow diagrams

Inbreeding coefficient of individual X pathways:

$$X \leftarrow S \xleftarrow{1} 4 \xleftarrow{2} 1 \xrightarrow{3} D \longrightarrow X = \left(\frac{1}{2}\right)^3 = 0.1250$$

$$X \leftarrow S \xleftarrow{1} 1 \xrightarrow{2} D \longrightarrow X = \left(\frac{1}{2}\right)^2 = 0.2500$$

Total = 0.3750

Thus $F_X = \frac{1}{2}(0.3750) = 0.1875 = 18.7\%$

Relationship coefficient between X and 1 pathways

Thus $R_{XI} = 0.6250 \sqrt{\dfrac{1+0}{1+0.1875}}$

$= 0.6250 \sqrt{\dfrac{1}{1.1875}}$

= 0.6250 √0.8421
= 0.6250 × 0.9177
= 0.5736
= 57.36 %

Problems

Calculate relationships coefficient between single first causins (X & Y) from the following pedigree and arrow diagram

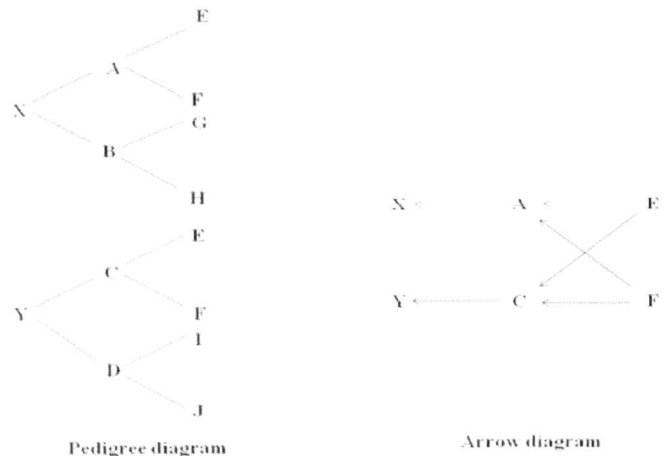

Pedigree diagram   Arrow diagram

Calculate relationship, coefficient between double first cousins (X and Y) from the following pedigree and arrow diagram

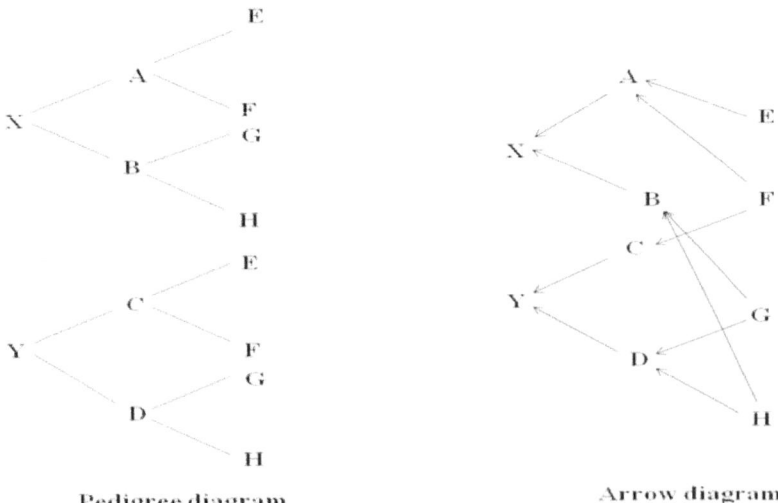

Pedigree diagram   Arrow diagram

Calculate relationship coefficient between son (X) and sirs (S) from

the follwomg pedigree and arrow diagram

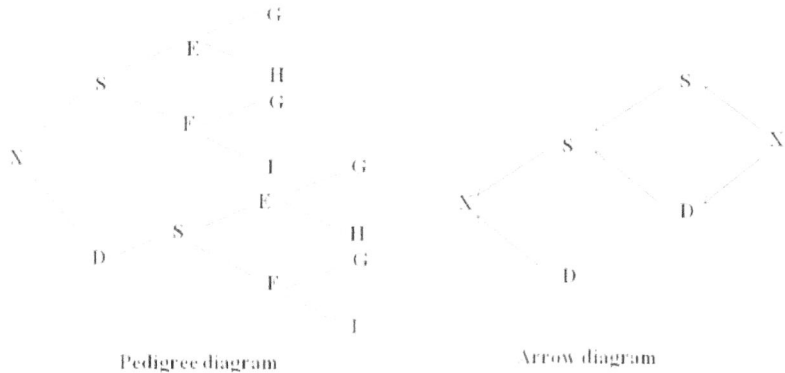

Pedigree diagram   Arrow diagram

C. Breeding for increased heterozygosity (out breeding)

A breeding system in which matings are between animals less closely related than the average of the group to which they belong i.e. the mates have no ancestors in common in the first 4-6 generations of their pedigree. Thus outbreeding is opposite of inbreeding. Examples of out breeding are mating between inbred lines. Strains or individuals within the same breed (out crossing). Between breed (cross breeding), extreme crosses between animals belonging to different species (Ass x Horse – mule).

Two Way Crosses

Two different populations (inbred lines, strains or breeds) are crossed with each other to produce an F_1 generation which exhibit hybrid vigor especially when crosses involve lines. The individuals of the F_1 generation are used for commercial purposes and not breeding. When two inbred lines of the same breed are crossed, the progeny are said to be in-crossbred.

Three – way crossed

The first generation crossbred females are crossed with males of a third line, strain or breed, thus utilizing the hybrid vigor of the dams. This is specially important in the crossing of inbred lines.

Four – Way or Double Crossed

Populations A and B are first crossed with each other, and so are C and D, to obtain the F_1 generations F_{AB} and F_{CD}. These are then crossed together to give the double hybrids, $F_{(AB.CD)}$. This method is used extensively in poultry breeding for crosses between inbred lines which have low viability.

Back – Cross

Usually the F_1 females are backcrossed to males of one of the parental populations e.g. combination $F_{(AB.A)}$ for the production of commercial animals. This method is advantageous especially when the F_1 females, on account of their hybrid vigor, are better mothers than females from either of the parent populations (material influence) especially in pig breeding. However, for economic reasons rotational crosses are preferred over back crosses.

Rotational Cross Breeding

Crossbred females are mated with males from either of the parent populations with the provision that the matings are alternated for each new generation. This system is mostly used in pig breeding. The following types of rotational crossbreeding have been tried.

Criss Crossing

Breeds A and B are crossed to produce an F_1 generation, the F_{AB} sows are backcrossed to boars from breed B, the $F_{(AB.B)}$ sows are then mated back to board from breed 4 and so on. The system has advantage over the two way cross, that one can continue to use the crossbred sows for breeding and it is only necessary to purchase purebred boars. Criss crossing system soon come to have about $\frac{2}{3}$ of their inheritance from the breed of their immediate sire and $\frac{1}{3}$ from the other breed being used in the system. The highest proportion of inheritance can be obtained from the formula $\frac{(50 \times 2)^n}{2^n - 1}$ where n = number of breed used in the system.

Scheme of criss crossing

1st Year	2nd Year	3rd Year	4th Year	5th Year
Boar bread	Boar bread I	Boar bread II	Boar bread I	Boar bread II
Crossbred sows	Crossbred sows	Crossbred sows	Crossbred sows	Crossbred sows
Breed II 50%	Breed II 25%	Breed II 62.5%	Breed II 31.25%	Breed II 65.375%
Breed I 50%	Breed I 75%	Breed I 37.5%	Breed I 68.75%	Breed I 34.625%

Three – Way Rotation

F_{AB} females are mated with boar of a third breed C. Boars from breed A are used on the next generation of females and board from breed B on the following generations of females and so on. The crossbred soon come to have about $\frac{4}{7}$ of the inheritance of the breed of their immediate sire, $\frac{2}{7}$ from the breed of their maternal grand sire and $\frac{1}{7}$ of the other breed being used. This system is also used in cattle and poultry breeding.

Four – Way Rotation

Males from a fourth breed D are used on females of the combination $F_{(AB.C)}$. Thereafter males from breeds A, B, C and D are used in succession for each new generation.

Top Crossing

This term is used for the mating of inbred males of females of non-inbred populations. It is questionable whether the greater homozygosity of the males for production traits compensates for their lower fertility and vigor.

Crossing with Recurrent Reciprocal Selection (RRS)

In each of two breeds or strains, progeny testing is carried out by crossing with the other breed or strain (reciprocal crossing). Those animals which produce the best progeny from such crosses are then used for multiplying their own breed or strain. The object is to change both populations gradually, so that they give better results in crosses with each other.

Grading up

Back-crossing to the same breed takes place generation after generation. The object is to change a mixed population to a 'pure breed'. With each new generation, the proportion of genes from the original mixed population decreases to half the proportion present in the previous generation, so that after 4 generations it will decrease to 6.25% and after 5 generations to 3.125%. This system of mating has been widely used in many countries where it was desired to change cattle of mixed or undescript ancestry to a particular recognized breed type. In tropical and sub-tropical areas where the purebred of the temperate zone in grading programmes are not adopted, often it has been observed that the pure breds degenerate when used in tropical and sub-tropical areas and their offsprings with higher inheritance (75% and more) of exotic blood do not have the vigor and constitution for high production and are susceptible to deseases, management and feeding conditions of the warmer climate and the system has not been very successful.

Crossing for the production of a new breed

The great majority of our present day breeds of farm animals have been founded by crossing different breed types with each other in an attempt to combine their desirable traits in new breed. The consolidation of a new breed, following the foundation crosses, generally demands a certain amount of inbreeding combined with intensive selection.

The breeding system 8 and 9 are concerned with increasing the heterozygosity of the population only in the initial stages. With continued crossing the heterozygosity will generally begin to decline again.

Examples of New Breed Developed by Crossbreeding,

Dairy Breed:

Jamacia Hope (Jersey 80%, Sahiwal 15%)

Jersindh (Jersey, Red Sindhi)

Tailor (Local, Jersey, Shorthorn)

Karan Swiss (Brown Swiss, Sahiwal + Sindhi)

Holstein Friesian (Holstein, Friesian)

Karanfries (Holstein Friesian x Tharparkar)

Jarthar (Jersey x Tharparkar)

Sunandani (Brown Swiss x local breed of Kerala)

Frieswal (Holstein Friesian x Sahiwal)

Beef Breeds:

Santa Gertrudis (Brahman 37.5%, Shorthorn 62.5%)

Beef master (Shorthorn, Hereford, Brahman)

Brangus (Brahman, Anngus)

Charbray (Brahman, French Charollais)

Swine Breeds:

Hamprace (Hampshire, Landrace)

Minnesota No. 1 (Tamworth, Landrace)
Minnesota No. 2 (Ploand China, Yorkshire)
Beltsville No. (Landrace, Poland China)
Beltsville No. 2 (Landrace, Hampshire, Duroc, Yorkshire)
Examples of species hybridication:
Mule (Jack of Ass species x mare of Horse species)
Hinny (Reciprocal of mule)
Zebroid (Zebra x Horse)
Pien Niu (Cattle x Yak in Tibet)
Cattalo (Bison bulls x Domestic Cows)
Genetic basis of outbreeding:
Out breeding would result in:
Covering up of undesirable recessive and

Increase in heterozygosity resulting into heterosis. Positive heterosis would result into icreased performance which would be most marked in traits depressed by inbreeding viz. reproductive traits – fertility, fecundity, maternal qualities etc. greatest increase in performance would be expected from crosses having greatest diversity of origin.

Selection For Combining Ability

1. Selection for General Combining Ability

For measuring the general combining ability, top crossing is followed. In top crossing, individuals from the inbred lines to be tested are crossed with individuals from the base population. The mean value of the progeny measures the general combining ability of the line because the gametes of individuals from the base population are genetically equivalent to the gametes of a random set of inbred lines derived without selection from the base population. This method is for comparing the general combining abilities of different lines and to choose the lines most likely to yield the best cross among all the crosses that would be made between the available lines.

2. Selection for Specific Combining Ability

Selection for specific combining ability means that selection is practiced to take advantage of hybrid vigour when non - additive gene action is important. In general, the heterogenous individual does not breed true. Heterozygous parents, produce approximately the half heterozygous offspring., the other one half being either homozygous dominant or recessive.

Selection For General and Specific Combining Ability

The specific combining ability of a cross cannot be measured without making and testing that particular cross. To get SCA, two lines should be developed which differ in gene frequencies. Two methods of selection are available viz., recurrent selection and reciprocal recurrent selection.

Recurrent Selection

In this system, a highly inbred line, presumably homozygous at most loci is selected as a "tester". A large number of individuals are tested in crosses with this line. The individuals giving the best results are inter-mated and a large number of their progeny again tested on the inbred tester. The cycle is repeated over and over. If heterosis is largely dependent upon overdominance, this would result in the line selected on cross performance becoming homozygous for different alleles than the inbred used as the tester.

Reciprocal Recurrent Selection

In this system, randomly selected representatives of each of two lines are progeny tested in crosses with the other. Those individuals of each line having the best cross progeny are then inter-mated to propagate their respective lines. Offsprings from these within-line-matings are again progeny tested in cross with the other and the cycle are repeated. The system should lead to improved cross performance whether it is the result of overdominance, dominance, epistasis, or only additive effects.

Both these systems involve progeny testing. Due to the increased generation intervals, this would be expected to result in slower progress than other breeding systems for characters moderate to high in heritability. They would be expected to be more useful than other breeding systems only if overdominance or other non-additive types of inter- or intra-allelic gene action are important in heterosis.

Chapter - 19

Heterosis or Hybrid Vigor

Heterosis, hybrid vigor, or outbreeding enhancement, is the improved or increased function of any biological quality in a hybrid offspring. The adjective derived from heterosis is heterotic. Heterosis is the occurrence of a superior offspring from mixing the genetic contributions of its parents. These effects can be due to Mendelian or non-Mendelian inheritance.

Definitions

In proposing the term heterosis to replace the older term heterozygosis, G.H. Shull aimed to avoid limiting the term to the effects that can be explained by heterozygosity in Mendelian inheritance.

The physiological vigor of an organism as manifested in its rapidity of growth, its height and general robustness, is positively correlated with the degree of dissimilarity in the gametes by whose union the organism was formed. The more numerous the differences between the uniting gametes — at least within certain limits — the greater on the whole is the amount of stimulation. These differences need not be Mendelian in their inheritance. To avoid the implication that all the genotypic differences which stimulate cell-division, growth and other physiological activities of an organism are Mendelian in their inheritance and also to gain brevity of expression. I suggest that the word 'heterosis' be adopted.

Heterosis is the opposite of inbreeding depression. Inbreeding depression leads to offspring with deleterious traits due to homozygosity. The term heterosis often causes controversy, particularly in selective breeding of domestic animals, because it is sometimes claimed that all crossbred plants and animals are genetically superior to their parents. This is untrue, as only some hybrids are genetically superior. The inverse of heterosis, when a hybrid inherits traits from its parents that are not fully compatible, with deleterious results, is outbreeding depression.

The term heterosis or hybrid vigour characterizes the increased ability or a hybrid as compared to the parental forms. The term ability means the favourable changes in hybrid characters when compared to the abilities of P (Parental) in one or several characters. This phenomenon was discovered in the 18th century by Veldeiter during experiments on tobacco hybridization and later confirmed by many selectionists working on hybridization of plants and animals. It was noticed that the main value of inbreeding - retention of desired characters - is connected to an equal degree with the risk of their diminution. Darwin (1876) was the first to attempt to explain the theoretical fundamentals of heterosis. He concluded that cross-pollination usually produced a favourable effect and spontaneous pollination a detrimental one. Darwin's works on comparison of spontaneous pollination and cross-pollinated plants gave rise to many investigations on selection. By the time of the re-discovery of Mendel's laws, considerable practical material was collected which confirmed the conclusion made by Darwin and also the two very important earlier observations made by Kolreuter (1763,1766) (hybrid ability is connected with the degree of genetic difference between their parents; hybrid vigour is of special importance in the process of evolution).

Theory of Genetic Balance

The theory of genetic balance formulated by Turbin (1966) attempts to coordinate the ideas available into a general theory of heterosis. The theory of genetic balance proceeds from the concept that if the development of a character is the result of genetic balance (which is the equilibrium attained between oppositely directed actions of different genes on the given character), the removal, modification or substitution of some of them will necessarily give advantage to the factors of opposite action. In organisms with broken genetic balance it may cause change in the extent of development of some other characters. Proceeding from this concept, heterosis may be developed as one of the consequences of a modified genetic balance in hybrids produced from non-related lines. It is necessary to note that sterility of genetically non-balanced forms in some cases does not exclude their practical application (only F_1 is used).

As can be seen, the theory of genetic balance provides a too general explanation of the causes of heterosis; but it does not specify the role of some other types of interaction of heredity factors which determine the phenomenon, and which are the components of genetic balance of hybrids. This theory is as correct as any other to explain the general notions, but at present it is hardly plausible as a general theory of heterosis.

Practical Application of Heterosis

The phenomenon of heterosis has been observed in all the species studied. In many cases, heterosis is so obvious that there is no need to resort to statistical analysis to demonstrate its value. This is particularly true of corn hybrids.

Methods of Inducing Heterosis

In selection of lines, their qualities are estimated in connection with the properties which should be obtained in a future hybrid organism. Inbreeding cannot be effective if not accompanied by selection. Having created a great number of lines, crossing is begun. Interlinear hybrids of the first generation are estimated by a heterosis effect; proceeding from this, the lines of better combining ability are selected and then reproduced at a greater rate for production of hybrid stock. At selection stations, work on production of inbred lines and estimation of their combining ability is carried out continuously. Crossing of inbred lines from one or various breeds has been widely used now in the field of poultry and pig breeding. It is necessary to note that wide scale utilization of hybrids in cattle breeding is possible only at the highest level of pure-strain stock farming where there is an availability of valuable breeds.

It is clear that in most cases the inbred lines will always have lower indexes than the strains. Heterosis is evident only when the interlinear hybrid exceeds not only its parent lines but also the varieties or breeds from which these lines generate.

Genetic nature of inbreeding in this case is the process of segregation of population in line with different genotypes.

Combining ability of a line or a species is heterosis development in hybrids obtained from their crossing.

Combining Ability

In crossing aimed at obtaining heterosis effect particular attention should be paid to combining capacity, distinguishing between general and specific combining ability. The first is characterized by an average amount of heterosis, observed in all hybrid combinations; the second, by deviation from this amount in one or another separate combination. For determination of general and specific combining ability, the form under test and the corresponding analyses (tester) are crossed. Moreover, for determination of general combining ability it is better to use analyses with a wide genetic basis (random pollination grade or corresponding animal population). The specific combining ability of the form being tested is estimated in relation to any form with which it is to be crossed later. This second form is an analysis. Choice of analyses depends on the purpose of the lines tested, i.e. whether they will be used to substitute the line in the existing hybrid combinations, or for the production of new hybrids.

The analysis for determination of combining ability of the material being selected is used for a system of crossings. For this purpose various systems of crossing are used, which are in fact various methods of determination of combining ability.

In plant growing, the following four methods are used: diallelic crossing, topcross, poly-cross, and random pollination. Of these methods only top-cross can be used in cattle breeding.

It is evident that the combining ability (general and specific) can be improved as a result of selection of recombinant forms having new combinations of genes. In this way it is possible to obtain new genotypes of a higher combining ability. Hybridization and selection are the means with which favourable complexes of hereditary factors are concentrated in a population. Effectiveness of the above means, in practice, that selection depends to a great extent on the methods applied for detection of genotypes which differ in combining ability. The complete programme of selection for determination of combining ability, the purpose of which is to obtain new components for crossing or for improving the existing lines, must provide for certain alternations of ways to produce new genetic combinations, their breeding estimation for combining ability, and selections of the best genotypes.

Depending on the direction of selective breeding of new lines, improvement of existing lines, or selection for determination of general or specific combining capacity, there are various selection programmes in existence. Various kinds of periodic selection are used, as well as convergent improvement, cumulative selection and gamete selection.

Periodic Selection

Periodic selection provides for alternation in crossing of the material selected with the inbred material and selection in the intermediate period. The selection is carried out on the basis of estimation of species selected by the characters which are of interest.

Convergent Improvement

This is a method suggested for corn selection. The theoretical basis of the method involves the fact that heterosis is considered as the result of a favourable effect of dominant linkage factors. By this method, using the back-cross of common hybrid A × B for the parental lines, new self-pollinated lines A' and B' are obtained, each of them carrying a number of genes of the second parental line. In this way the accumulation of favourable dominant genes in new parental lines A' and B' is ensured.

Cumulative Selection

Cumulative selection is also based on the theory of dominance. This method is used to accumulate favourable dominant genes by crossing the lines which have excellent combining ability, and selecting the best combinations.

Gamete Selection

Theoretically this is related to the phenomenon of superdominance. This method provides for improvement of the selected self-pollinated line on the basis of recombination, achieved by random selection of gametes of the strain. In order to obtain a new improved line, the line selected is crossed with a definite strain or hybrid, assuming that the offspring will differ from the initial line only by such properties (or gametes) which were introduced by the strains.

Retention of Heterosis

The problem of hybrid vigour retention, i.e. constant development of heterosis is a sequence of generations, has several theoretical approaches, some of which are applied in practice in plant growing.

However, in animal breeding, the problem of retention of hybrid vigour remains to a considerable degree unsolved.The theoretical basis of heterosis retention is accumulation in a population of balanced, closely-linked gene blocks, the combination of which in the heterozygote leads to heterosis. In other words, it is necessary to avoid segregation of the valuable genotype obtained as a result of crossing.

One of the approved methods of retention of heterosis is asexual reproduction of hybrids. This method of retention of hybrid vigour is widely used in selection of fruit plants.

Another form of asexual reproduction in plants is apomixis. In some cases of apomixis the seeds develop from diploid cells, both stages of meiosis being omitted, and the generation of these seeds is genotypically identical to the mother plant. Retention of heterosis through apomixis is used in growing blackberry, roses, meadow grass, citrus and many other plants.

Retention of hereditary heterosis may occur as a result of combination of lethal genes and certain chromosomic arrangements. In a specimen, heterozygous by inversion, the crossing-over inside of an inverted section leads to development of sterile gametes. As the crossovers obtained inside of an inverted section are suppressed due to sterility of crossover gametes, the heterozygosis of the inverted section becomes permanent. This type of balanced lethality is observed in hymenopteran insects.Besides, heterozygosis can be retained or improved by means of a definite combination of recessive lethal genes with other types of chromosomic arrangements - translocations. Retention of hybrid vigour sometimes is obtained with the aid of polyploidy. The practice of plant culture shows that increased vigour of interspecific hybrids is observed, not only in diploid hybrids which can be sterile, but also in amphidiploid forms, in which the chromosomic combination is doubled. Interspecific hybrids with a double combination of chromosomes are very often fertile, and their descendants continue manifesting heterosis developed as a result of the genotypes from two species.

Especially of interest in fisheries is a very promising method of retention of heterosis with the help of gynogenesis. It has been proved that in some cases the gynogenetic development of the zygote in fishes results in the emergence of clones, i.e. forms genotypically absolutely identical to the mother organism. In other words, meiotic segregation is absent in this case. Thus, development of gynogenesis in fishes on a commercial scale might facilitate the solution of the problem of retention of heterosis.

Hybrid Livestock

The concept of heterosis is also applied in the production of commercial livestock. In cattle, hybrids between Black Angus and Hereford produce a hybrid known as a "Black Baldy". In swine, "blue butts" are produced by the cross of Hampshire and Yorkshire. Other, more exotic hybrids such as "beefalo" are also used for specialty markets.

Within poultry, sex-linked genes have been used to create hybrids in which males and females can be sorted at one day old by color. Specific genes used for this are genes for barring and wing feather growth. Crosses of this sort create what are sold as Black Sex-links, Red Sex-links, and various other crosses that are known by trade names.

Commercial broilers are produced by crossing different strains of White Rocks and White Cornish, the Cornish providing a large frame and the Rocks providing the fast rate of gain. The hybrid vigor produced allows the production of uniform birds with a marketable carcass at 6–9 weeks of age.

Likewise, hybrids between different strains of White Leghorn are used to produce laying flocks that provide the majority of white eggs for sale in the United States.

Two approaches for heterosis estimation on 3-way cross

The first approach for calculating the 3-way heterosis is directly using the widely-used empirical formula. The widely-used empirical formula for 3-way heterosis estimation in the Chinese pig industry, denoted as Ax (BxC), is generally described as follow:

$$H\% = \frac{\bar{F} - (0.5\bar{A} + 0.25\bar{B} + 0.25\bar{C})}{0.5\bar{A} + 0.25\bar{B} + 0.25\bar{C}} \times 100 \qquad (1)$$

where, H, % is the value of heterosis ratio (the numerator is heterosis), $\bar{F}$ is the phenotypic mean of triple cross Ax (B x C) and $\bar{A}$, $\bar{B}$ and $\bar{C}$ are the phenotypic means of parental lines A, B and C or breeds A, B and C, respectively.

The second approach is the Cockerham's model-based estimation approach. Considering the widely-reported use of Cockerham's model in heterosis studies (Yang, 2004; Kao and Zeng, 2002; Xu and Zhu, 1999; Cockerham and Zeng, 1996), we here use the Cockerham's model to conduct the comparative estimation. For an arbitrary animal, the phenotype can be partitioned as follow:

$$y = \mu + G + e \tag{2}$$

where

- y = The phenotype
- μ = The mean value of population
- G = The genetic effect (hereinto, $G = A + D + I$ in which A is the additive effect, D is the dominant effect and I is the epistatic effect)
- e = Comprises systematical effect E under multi-environmental conditions

For simplification, we ignored the epistatic effect and assumed the inheritance of trait is followed an additive-dominance model, as done in other researches (Xu and Zhu, 1999) and the component G can be partitioned according to the Cockerham's model as follow:

$$G = \sum_i \alpha_i A_i + \sum_i \sum_j \delta_{ij} D_{ij} \tag{3}$$

where, α_i is the coefficient of additive genetic effect and $\Sigma_i \alpha_i = 2$ and δ_{ij} is the coefficient of dominance genetic effect and $\Sigma_i \Sigma_j \alpha_{ij} = 1$. Thus, for the 3-way hybrid, Ax (BxC), the genetic components of straightbred A, B, C and their crossbreds can be partitioned as follows:

$$G_A = 2A_A + D_{AA} \tag{4}$$

$$G_B = 2A_B + D_{BB} \tag{5}$$

$$G_C = 2A_C + D_{CC} \tag{6}$$

$$G_{BC} = A_B + A_C + D_{BC} \tag{7}$$

$$G_{A(BC)} = A_A + \frac{1}{2}A_B + \frac{1}{2}A_C + \frac{1}{2}D_{AB} + \frac{1}{2}D_{AC} \tag{8}$$

On the biological meaning of heterosis that is biologically determined by the gap between the performance of hybrid progenies and the average performance of their parents, the genetic components of heterosis of triple-cross population can be estimated as follow:

$$H_M = G_{ABC} - \frac{1}{2}(G_A - G_{BC}) \qquad (9)$$
$$= A_A - \frac{1}{2}A_B + \frac{1}{2}A_C - \frac{1}{2}D_{AB} - \frac{1}{2}D_{AC} - A_A -$$
$$\frac{1}{2}D_{AA} - \frac{1}{2}A_B - \frac{1}{2}A_C - \frac{1}{2}D_{BC}$$
$$= \frac{1}{2}D_{AB} + \frac{1}{2}D_{AC} - \frac{1}{2}D_{AA} - \frac{1}{2}D_{BC}$$

where, H_M is the heterosis of triple-cross progeny according to the biological definition of heterosis.

Let's consider the bias of the empirical formula from the Cockerham's model-based approach. Following the equational structure of the empirical formula, the genetic components of the empirical formula-based 3-way heterosis (1) can be partitioned as follow:

$$H = \frac{1}{2}D_{AB} - \frac{1}{2}D_{AC} - \frac{1}{2}D_{AA} - \qquad (10)$$
$$\frac{1}{4}D_{BB} - \frac{1}{4}D_{CC}$$

Here, it is clear that the expectation of the between H_M and H is $1/2\ D_{BC}$ -$1/4\ D_{BB}$ -$1/4\ D_{CC}$, which is just the half of the value of heterosis in F_1 single-cross generation, namely the theoretical genetic bias of the empirical formula is the half of the heterosis effects of F_1 single-cross.

Heterosis Definitions : Heterosis is a measure of the superior performance ofthe crossbred relative to the average of the purebredsinvolved in the cross. The probable cause of most heterosis is due to combining genes from different breeds, concealing the effects of inferior genes. Heterosis may resultin the crossbred being better than either parental breedor simply better than the average of the two. For example, an Angus x Hereford crossbred calf may generallygrow faster than either Angus or Hereford purebreds.However, a Charolais x Angus crossbred calf may notgrow as fast as a purebred Charolais or have the abilityto marble similar to purebred Angus but will likely be better than the average of the purebreds for both traits.This difference is usually expressed as a percentage of the average performance of the purebreds. The general formula for calculating percent heterosis isgiven below:

% heterosis = <u>crossbred avg. - purebred avg.</u> x 100
                          purebred avg.

Example 1 illustrates the calculation of the percentage of heterosis.

Average breed performance for weaning weight is given for two arbitrary breeds. In this example, the heterosis value of 4.4% means that the crossbred progeny performance is 4.4% greater than the average parental breed performance.

1. Heterosis for Weaning Weight Breed A: 455 lb. Breed B: 445 lb.

Purebred Average = (455+445)/2 = 450 lb.

Crossbred Average = 470 lb.

% heterosis = $\frac{470 - 450}{450}$ x 100 = 4.4%

Types of Heterosis

Heterosis arises from three mating situations. Individual heterosis is the advantage of the crossbred individual relative to the average of the purebred individuals. For example, a Limousin x Hereford calf may grow faster than the average of purebred Limousin and Hereford. Maternal heterosis is the advantage of the crossbred mother over the average of purebred mothers. For example a Hereford x Angus cow is generally a better mother (higher weaning %, milk production, etc.) than the average of purebred Hereford and Angus dams. Paternal heterosis is the advantage of a crossbred sire over the average of purebred sires. Paternal heterosis generally has an effect only on conception rate and aspects of male reproduction. The male parent does not have any direct environmental effect on the survival of the calf, so the beneûts are more limited than those for maternal heterosis. However, the beneût in added conception rate can be substantial, particularly if young males are being used. Numerous experiments have been conducted to investigate the effects of crossbreeding in cattle. These experiments yielded estimates of heterosis as well as comparisons among the breeds involved. Table 1 summarizes results on the percentage of heterosis for various traits based on research from several locations.

Heterosis may be a positive or negative value. Also, heterosis may be positive even when one of the parent breeds outperforms the crossbred average. Heterosis values (Table 1) have been derived largely from experiments involving British and/or European breeds. Note that British crosses with Brahman and other Bos Indicus breed types may result in more heterosis due to less genetic similarity between breeds. Not all traits express the same degree of heterosis. Heterosis levels can be grouped into three major classes. Reproductive traits generally show fairly high levelsof heterosis. Growth traits generally have moderate levels of heterosis while carcass traits infrequently displaymuch heterosis. There are exceptions to these generalities but the three classes work as a general rule ofthumb. It should be pointed out that this is exactly thereverse of the general levels of heritability for theseclasses of traits. For example, carcass traits show lowlevels of heterosis; however, these traits tend to be highly heritable. Also, traits that express the higher levels of heterosis tend to be those traits that are more highlyinûuenced by inbreeding. This should make sense asheritability depends upon additive types of gene actionwhile heterosis and inbreeding depression depend uponnon-additive gene action such as dominance and epistasis. Heterosis also requires genetic differences amongindividuals involved in the crosses.

Use of Heterosis

Heterosis levels are presented as percentage values sothey can be used to calculate the expected performanceof the crossbred individuals. The ûrst step calculates theaverage expected performance of the purebreds. This canbe accomplished, simplistically, by multiplying eachbreed value by the proportion it contributes to the cross:

Two-Breed Cross: Hereford x Angus

Expected crossbred performance = 1D 2 H + 1D 2 A + heterosis

The heterosis to be added into a two-breed cross isindividual heterosis. This is accomplished by multiplying the purebred average by the % heterosis and adding it to the average of the breeds. If the dam is also crossbred, the maternal heterosis would have to be included, in the same manner as the individual heterosis. Example 2 shows the procedures involved.

Example 2. Calculating Expected Weaning Weight of CrossbredCalves Individual Heterosis: 4.7 % .Maternal Heterosis: 4.2 %.Two Breed Cross (purebred A sire x purebred B dam)Breed A: 460 lbBreed B: 480 lbPurebred Average = 1D 2 (460) + 1D 2 (480) = 470 lb

Individual Heterosis = .047(470) = 22.09 lb

Expected Crossbred Average = Purebred Average + Individual Heterosis

Expected Crossbred Performance = 470 + 22.09 = 492.09 lb.

Three-Breed Cross (purebred C sire x crossbred AB dam)

Breed C: 500 lb.

Purebred Avg. = 1D 2 (500) + 1D 2 (460) + 1D 2 (480) = 485 lb.

Expected Crossbred Avg = Purebred Avg + Individual Heterosis + Maternal Heterosis

Individual Heterosis = .047 (485) = 22.80 lb.

485 + 22.80 = 507.80 lb.

Maternal Heterosis = .042 (507.80) = 21.33 lb.

Expected Crossbred Performance = 507.8 + 21.33 = 529.13 lb.

Types of Dominance

This simple example assumed that cattle from the three breeds perform equally well as sires or dams. In other words, this case assumed no breed complementarity. This assumption makes the calculations simple but is not very useful in practical situations. For example, it is well established that a Charolais sire x Angus dam cross will perform differently than an Angus sire x Charolais dam cross. It is necessary to calculate expected performance such that the relative merit of the breeds as dams is taken into account. Substantial differences exist between breeds in reproductive performance and in mothering ability. To include maternal differences between breeds, each breed will be assigned two values. The ûrst value will be for its contribution of genes to the offspring (direct breed value). The second value will be for any superiority or inferiority as a dam (maternal breed value). The expected performance will then take on the following general form (with the general purebred mean representing the average performance of all the considered breeds when used as purebreds):

Expected performance = General Purebred Mean+ 1D 2 sire breed direct value + 1D 2 dam breed direct value+ dam breed maternal value + individual heterosis + maternal heterosis .The heterosis values will still be added on as increases after the breed values are added together. The direct values are multiplied by 1/2 because each parent contributes half the genes to the offspring.

The relative values of the direct and maternal effects will depend upon the biological mechanism for the traits. Some traits, such as post weaning average daily gain, are largely determined by the genotype of the individual. These traits will have relatively small contributions from maternal effects. Other traits, such as calving percentage, are essentially determined by the dam so they will have very large maternal effects relative to the direct effects. A trait like weaning weight has large direct components as well as maternal components so both values may be large in magnitude.

Loss of Heterosis

The examples presented thus far were for situations where individual and maternal heterosis were either completely absent or completely present. This is not always the case. Heterosis (from dominance) is dependent upon gene pairs having members from two different breeds. In backcrosses, rotational crossbreeding and composite breeds, not all gene pairs will have genes from different breeds. The proportion of available heterosis (arising from dominance) can be easily predicted by examining the degree of heterozygocity. In a typical two breed cross, all gene pairs will be crossbred (one gene from each parental breed) and will be more likely to be heterozygous or become of different breed sources: Hereford x Angus cross (with 8 pairs of genes)Hereford x Angus = Crossbred

|H| |H| |A| |A| |H| |A|
|H| |H| |A| |A| |H| |A|
|H| |H| |A| |A| |H| |A|
|H| |H| |A| |A| |H| |A|
|H| |H| |A| |A| |H| |A|
|H| |H| |A| |A| |H| |A|
|H| |H| |A| |A| |H| |A|
|H| |H| |A| |A| |H| |A|

This example illustrates one pair of chromosomes,(with eight genes on each chromosome) for each animal. The Hereford has only Hereford genes, and theAngus has only Angus genes. The crossbred hasHereford genes on one chromosome and Angus geneson the other chromosome so that each gene pair has4 Beef Cattle Hand book x =one gene from each breed. This calf is completely crossbred. If the crossbred calf is mated back to a Hereford bull (shown below), there will be crossing over and random segregation so that the egg cell produced by thecrossbred heifer will have half Hereford genes and halfAngus genes. The result of this backcross will have thefollowing genetic makeup:

Hereford x HA = Backcross

|H| |H| |H| |A| |H| |H|
|H| |H| |H| |A| |H| |H|
|H| |H| |H| |A| |H| |H|

|H| |H| |H| |A| |H| |H|
|H| |H| |H| |A| |H| |A|
|H| |H| |H| |A| |H| |A|
|H| |H| |H| |A| |H| |A|
|H| |H| |H| |A| |H| |A|

This calf's gene pairs are half H x A and half H x H. We might say that only half of the gene pairs are crossbred. As a result, only half of the individual heterosis is present. The dam is the result of a Hereford x Angus mating so all of the maternal heterosis is present. The amount of heterosis that can be utilized can be calculated by considering the proportion of gene pairs of an individual which are crossbred. The H x HA backcross had half crossbred gene pairs so 50 percent of the individual heterosis was used. The dam was completely crossbred, so 100 percent of the maternal heterosis was used. If heifers from the backcross were mated to a Charolais bull, there would be 100 percent individual heterosis since all gene pairs in the offspring would have one Charolais gene and one gene which is either Hereford or Angus. Any mating system that results in the sire breed(s) and dam breed(s) having some commonality will lose some heterosis. One point of confusion for many producers is the loss of heterosis when a composite breed is formed. There would be a loss of heterosis, as described in this section, because of backcross matings. However, once the breed is established and matings are among individuals with similar genetic makeup, the level of heterozygocity and the resulting heterosis should be constant. For example, if a two-breed composite is formed with equal representation of both breeds, the level of heterozygocity stabilizes at 50 percent. This would remain true until selection and inbreeding began to move the genetic makeup of the composite breed toward more homozygocity. The loss of heterosis described in this section is that part due to dominance. Epistasis may lead to increased loss of heterosis or decreased loss of heterosis depending upon the nature of the epistasis. Experimental reports of epistatic effects in beef cattle are not numerous, but the existing evidence suggests that it is of some importance. The practical effect of these epistatic effects is that prediction of heterosis level due to heterozygocity will be biased.

Genetic Basis of Heterosis

The genetic basis of heterosis is the opposite of the origin of inbreeding depression. Inbreeding tends to cause more gene pairs in an individual to be homozygous (theoffspring receives an identical gene from each parent). Incontrast, crossbreeding tends to cause more gene pairsto be heterozygous (the offspring receives differentgenes from its parents). This arises from the fact that different breeds tend to have high frequencies of differentgenes. Breeds that are genetically very different would tend to cause more heterozygocity and, as a result, more heterosis when crossed. Heterozygocity will result in better performance if there is non-additive gene action(dominance and epistasis) and the recessive alleleresults in inferior performance. Dominance is present ifthe heterozygous individual is not exactly intermediatebetween the two homozygotes. This would be analogousto the dominance relationship that results in the offspring of a mating between a black bull and a red cowbeing black, with the red masked. Various types of dominance are illustrated in Figure 1. Epistasis may also play a part in heterosis. Epistasis is the interaction betweendifferent loci.

The relationship between scurs and horns. If an individual is homozygous for the horned condition, it is horned,regardless of any presence or absence of the scur gene.The gene location (locus) that affects scurs can onlyexpress itself in an individual that has at least one polledgene at the locus that affects horns.The evidence for the relative importance of dominance and epistasis in beef cattle heterosis is not conclusive, but studies have shown dominance to be themajor factor for many traits.

Full understanding of the genetic basis of heterosis will depend upon:

1. Extension of knowledge of the physiology of gene action.

2. Physiological reactions resulting in heterosis.

Genetic basis put forward for heterosis are:

1. Due to covering up the effects of deleterious recessive genes.

2. Due to epistasis or inter allelic interaction. Due to over dominance.

Chapter - 20

Selection of Dairy Cattle & Buffaloes

General Selection Procedures for Dairy Breeds

Selection of Dairy Cows

Selecting a calf in calf show, a cow in cattle show by judging is an art. A dairy farmer should build up his own herd by breeding his own herd. Following guidelines will be useful for selection of a diary cow. Whenever an animal is purchased from cattle fair, it should be selected based upon its breed characters and milk producing.

History sheet or pedigree sheet which are generally maintained in organized farms reveals the complete history of animal. The maximum yield by dairy cows are noticed during the first five lactations. So generally selection should be carried out during First or Second lactation and that too are month after calving.?Three successive complete milkings has to be done and an average of it will give a fair idea regarding production by a particular animal.

A cow should allow anybody to milk, and should be docile.?It is better to purchase the animals during the months of October and November.?Maximum yield is noticed till 90 days after calving.

Breed Characteristics of High Yielding Dairy Cows

- Attractive individuality with feminity, vigour, harmonious blending of all parts, impressive style and carriage.
- Animal should have wedge shaped appearance of the body
- It should have bright eyes with lean neck

- The udder should be well attached to the abdomen
- The skin of the udder should have a good network of blood vessels
- All four quarters of the udder should be well demarcated with well placed teats.
- Selecting breeds for Commercial Dairy Farm - Suggestions
- Under Indian condition a commercial dairy farm should consist of minimum 20animals (10 cows, 10 buffaloes) this strength can easily go up to 100 animals in proportion of 50:50 or 40:60. After this however, you need to review your strength and market potential before you chose to go for expansion.?Middle class health-conscious Indian families prefer low fat milk for consumption as liquid milk. It is always better to go for a commercial farm of mixed type. Cross breed, cows and buffaloes kept in separate rows under one shed. Conduct a thorough study of the immediate market where you are planning to market your milk You can mix milk from both type of animals and sold as perneed of the market. Hotels and some general customers (can be around 30%) prefer pure buffalo milk. Hospitals, sanitariums prefer cow's milk.
- Selection of cow/buffalo breeds for commercial farmCows

Good quality cows are available in the market and it cost around Rs.1500 to Rs. 1800 per liter of milk production per day. (e.g. Cost of a cow producing 10liter of Milk per day will be between Rs. 16,000 to Rs.18,000). If proper care is given, cows breed regularly giving one calf every 13-14 monthinterval. They are more docile and can be handled easily. Good milk yielding cross breeds (Holstein and Jersey crosses) has well adapted to Indian climate? The fat percentage of cow's milk varies from 3-5.5% and is lower than Buffaloes.

Buffaloes

In India, we have good buffalo breeds like Murrah and Mehsana, which aresuitable for commercial dairy farm.?Buffalo milk has more demand for making butter and butter oil (Ghee), as fatpercentage in milk is higher than cow. Buffalo milk is also preferred for makingtea, a welcoming drink in common Indian household? Buffaloes can be maintained on more fibrous crop residues, hence scope for reducing feed cost.?Buffaloes largely mature late and give birth to calves at 16 to 18 months interval.Male calves fetch little value.?Buffaloes need cooling facility e.g. wallowing tank or showers / foggers with fan.

Selection of She-Buffaloes for Milk Production

When you purchase buffaloes for milk production we have to select healthy animalknown for economic milk production. We have to take following steps in selecting a dairy animal

Breed Characters

- Body confirmation
- Body weight
- Ancestors performance
- Reproduction capacity
- Health condition
- Age
- No. of lactations
- Past performance of the animal
- Free of chronic disease
- Cleanliness of teeth Legs and toes free of injuries
- Good eye site
- Whether animal is dry or lactating
- Date of delivery
- Month of pregnancy
- If non-pregnant, how many times it came in to heat
- Animal should follow owner's instructions
- The udder should be in good shape and easy to milk
- The animal should not have the following
- Poor growth
- Late maturity
- Not coming into heat
- Repeat breeder

- Long gap between two lactations
- Incurable chronic diseases
- Retained placenta
- Low milk production
- Unable to give milk without calf

Chapter - 21

Current Livestock and Poultry Breeding Programmes

Cattle Breeding Policies and Programmes in the Planning Process

It is necessary to see how the issues relating to Cattle Breeding have been dealt with in the successive Plan periods starting from the First Five Year Plan. It is seen that, while the issue of surplus cattle has been handled in different ways in the various Plan documents, some discussing this issue at great length and other remaining completely silent on the subject, other issues such as cattle development, genetic improvement and breeding etc., have also received varying degrees of emphasis in the different Plans.

First FiveYear Plan

The First Plan document speaks of the fact that the available feed could not adequately sustain the then existing bovine population and noted that, while there was a deficiency of good milch cows and working bullocks, there existed a surplus of useless or inefficient animals, and that this surplus was pressing upon the scanty fodder and feed resources. It was suggested that a programme for improvement of cattle should be launched, involving arrangements for production and use of adequate numbers of superior bulls of known parentage and productivitiy and elimination of inferior and unapproved bulls.

Key Village Scheme

It was envisaged that, under the Key Village Scheme, 600 centres would be set up in the Plan period, each centre with three or four villages having about 500 cows of over three years of age where maintenance of records of pedigree and milk production, feeding and disease control would receive full attention and techniques of artificial insemination would be utilised by setting up one AI centre for four key villages centres. Improvement of common grazing grounds, growing of fodder crops in suitable rotations, preservation of surplus monsoon grass, and use of untapped fodder resources were some of the key components of the Scheme.

Second Five Year Plan

The document for the Second Five Year Plan notes that the object of animal husbandry programmes is, inter alia, to increase the supply of milk, meat and eggs and to provide efficient bullock power for agricultural operations in every part of the country, which meant that the quality of the cattle was of critical importance to the rural economy.

Cattle Breeding Policy and Programme

The Plan paper documents the fact that there are as many as 25 well-defined breeds of cattle and six well-defined breeds of buffaloes in India, which are distributed in different parts of the country. High class specimens in each breed are limited in number and are found in the interior of its particular breeding tract, around which there are animals of the same type but of poorer quality. A few of these breeds are of the dairy type while a large majority of the breeds are of the draught type. In between there are "dual-purpose" breeds, whose females yield more, than an average quantity of milk, while the males are good working bullocks. It was found that the while the well-defined breeds are largely found in the dry parts of the country, over large parts of the country in the east and the south of India where rainfall is very heavy, the cattle are non-descript and do not belong to any definite breed.

The major guidelines of the all-India breeding policy, drawn up by the Indian Council of Agricultural Research and accepted by the Central and State Governments were:

In the case of well-defined milch breeds the milking capacity should be developed to the maximum by selective breeding and the male progeny should be used for the development of the nondescript cattle.

In the case of well-defined draught breeds, the objective is to put as much milk-producing capability in them as possible, without materially impairing their quality for work. Thus, the breeding policy was generally designed to increase the production of milk in the country, without affecting the position in regard to the supply of bullocks required for cultivation. In every draught breed there is always a small number which give more than an average quantity of milk and by selecting bulls from this group, the milk production of the population could be progressively increased by further selection and breeding. When this is done in the interior of the breeding tracts, the bulls produced can be used in the outer areas in order that general improvement may be brought about in the entire population.

For the implementation of this policy, each State was divided into zones according to the breeds used in them. Thus, in the districts of Ahmedabad, Kaira, Broach and Surat the breed to be used was 'Kankrej'. In the western tracts of U.P. like Saharanpur, Muzaffamagar, Aligarh, Mathura, etc., the breed proposed to be used was 'Hariana'. In the hilly tracts such as Dehra Dun, Garhwal, Almora and parts of Nainital, where the cattle are non-descript, Sindhi bulls were to be used.

Key Village Scheme

It was envisaged that, mainly through the key village scheme, the programme of livestock improvement would be pursued by State Governments. This scheme provides for concentrated work in selected areas. It envisages castration of scrub bulls, breeding operations controlled by artificial insemination centres (each of which is intended to serve about 5,000 cows of breeding age), rearing of calves on a subsidised basis, development of fodder resources and the marketing of animal husbandry products organised on co-operative lines. During the first five year plan, 600 key villages and 150 artificial insemination centres had been established. During the second plan 1258 key villages, 245 artificial insemination centres and 254 extension centres were to be set up. The programme was intended to produce about 22,000 improved stud bulls, 950,000 improved bullocks and a million improved cows. The scheme made encouraging progress, but in respect of fodder development and the marketing of animal husbandry products not much headway was made. On the other hand, controlled breeding had found a large measure of acceptance and States had enacted the necessary legislation for implementing the scheme. In the early stages, work in many key villages and artificial insemination centres was delayed for want of equipment and shortage of staff, but everywhere the local people were willing to provide rent-free buildings and contribute in other ways to make the scheme a success. During the second plan, a great deal of attention was to be given to the fodder programme as this was an essential basis for the programme of cattle development. In each area efforts were to be made to develop the limited pasture lands which were available. With the large programme envisaged in the second plan, a high degree of urgency was atlached to the provision of adequate staff, to better administrative planning of supplies and to public education in matters affecting animal husbandrydevelopment.

Third Five Year Plan

The Third Five Year Plan document took note of the seriousness of the problem of surplus and uneconomic cattle and arrived at the conclusion that weeding of inferior stock was a necessary complement to a programme of cattle improvement and systematic breeding. The Plan states that having regard to the size of the problem of surplus cattle and its special features, with a view to elimination of scrub male stock, it was proposed to undertake a large-scale programme of castration during the Third Plan. The programme envisaged that mass castration work would be initiated first, in areas in which intensive livestock development programmes have been taken up and would be later extended to other areas.

Fourth Five Year Plan

This Plan has not specifically discussed the problem of surplus cattle but has talked about the cattle development programmes launched in the previous Plan period. It was proposed that the schemes of the Third Plan including those relating to cattle breeding farms, bull rearing farms, Goshala development, and control of wild and stray cattle and organisation of mass castration would continue and three central cattle breeding farms and eight bull rearing farms would be set up during the Fourth Plan period. It was also indicated that Sire-evaluation cells would be established in each State.

Fifth Five Year Plan

The document for the Fifth Five Year Plan has not mentioned the animal husbandry sector, and while discussing the perspectives on agriculture has singularly concentrated on food-grain production and related issues. Only in the Chapter on Plan Outlays and Programmes of Development, a small paragraph on Animal Husbandry and Dairy Farming find its place. Here it has been acknowledged that there had been some delay in giving a start to the special livestock development programmes through small and marginal farmers and agricultural labourers. By and large, the targets under production oriented projects such as the intensive cattle development (ICD) projects were expected to be fully achieved. There were 85 subsidised projects for calf-rearing. It was envisaged that the emphasis would continue to be laid on cross-breeding of cattle through establishment of exotic cattle breeding farms and intensive artificial insemination measures.

Sixth Five Year Plan

While reviewing the position with regard to animal husbandry and dairying, the Sixth Plan document notes that the increase in productivity of cattle and buffalo received continuing emphasis since the advent of the Planning process and progressive introduction of artificial insemination technique using superior breeding bulls was the main plank for cattle development under the Key Village Scheme and the Intensive Cattle Development programmes.

The Plan document noted that several special livestock production projects through small and marginal farmers and agricultural labourers were formulated based on the recommendations of the National Commission on Agriculture. Under this programme, 99 projects for subsidised rearing of cross-bred heifer calves were taken up in different States.

The document speaks of the need to increase the productivity of cattle by making concerted efforts to contain the increase in the population of cows and she buffaloes and to change the structure of these populations by replacing non-descript local stock by high-producing cows of indigenous breeds, cross bred cows and improved buffaloes. To achieve this, States were required to frame their breeding policies.

Seventh Five Year Plan

The Plan document for the Seventh Plan period speaks of the efforts to increase productivity of milch cattle in the previous Plan, through the establishment of 500 Key Villages and 122 Intensive Cattle Development projects. Cross-breeding with exotic dairy breeds was accelerated through the establishment of frozen semen stations in different States.

For increasing milk production and to improve draught power of bullocks, programmes for improvement of various breeds would continue, with emphasis on inputs like high merited breeding bulls, adequate and scientific feeding, modern management practices, provision of health facilities would continue and efforts would be made to bring at least 25 million cows under the cross-breeding programme.

Eighth Five Year Plan

In the Plan for the Eighth Five Year Plan, the need for paying special attention to technologies being developed to make activities in the livestock and dairy development sector economically more remunerative for the farmers. Emphasis was sought to be given to research in frontier areas such as genetic engineering which would provide for rapid upgradation of cattle through the use of Embryo Transfer Technology, development of more effective vaccines to control livestock diseasesandsoon.

Ninth Five Year Plan

The Ninth Plan paper documents a considerable improvement in production of milk during the previous Plan, which is attributed to the intensified activities particularly, in improvement of genetic stocks, through cross-breeding, effective control of diseases and the Operation Flood Programmes. The Ninth Plan sought to achieve the goals of doubling of food production and alleviation of hunger by adopting, for the first time a Regionally Differentiated Strategy based on the agro-economic and climatic conditions of different regions.

Animal Husbandry and Dairying, contributing about 26% of the total agricultural output was recognised as an important tool for generating employment and supplementing incomes of small and marginal farmers and agricultural labourers. The specific areas identified for intervention and support included, scientific management of genetic stock resources and upgradation, breeding, producing quality feed and fodder and so on.

3.10.3 One of the key research areas identified under Animal Sciences discipline was Genetic resource enhancement of cattle and other animals, through selection / cross breeding / embryo biotechnology.

Tenth Five Year Plan (2002-2007)

The major recommendations of the Report of the Working Group on Animal Husbandry set up by the Planning Commission for the Tenth Plan proposals, insofar as they relate to cattle developmentandbreedingpoliciesareasfollows:

A new programme focused exclusively on draught breeds of livestock may be initiated during Xth Plan.

The National Project for Cattle and Buffalo breeding may be continued and the stipulations made for its implementation may be followed in letter and spirit to realize the envisaged targets. It is necessary that adequate budgetary support be provided to this scheme to enable sequential development of the breeding networks in a given time frame.

A well defined livestock breeding policy is in place which states that pure indigenous well developed breeds should be improved through selection, while non-descript low producing populations should be improved by grading up with other superior indigenous breeds or crossbreeding with exotic males. It was observed that crossbreeding with exotic breeds is practiced even in home tract of elite important indigenous breeds. This is threatening the very existence of these breeds in their home tracts. It is recommended that the government should initiate steps to create incentives for breeding indigenous elite breeds and improve them through selection.

An aggressive strategy is to be adopted to remove the hurdles in sourcing and use of quality bulls for breeding. Military dairy farms could be used as a major source of crossbred bulls. They can give 5,000-7,000 crossbred bulls every year for the national bull production programme.

Monitoring cell for certification of sperm stations and A.I. bulls should be established in each state. Only certified semen should be used for A.I; where certification of semen is not possible, bulls may be used for breeding.

Institutional arrangements for production and delivery of breeding inputs may be reviewed and restructuring as required may be adopted on priority basis. Government may withdraw gradually from the production and delivery of breeding inputs and create a congenial environment and play a supportive role for private operators to grow. Government should recover the delivery and input cost of A.I service on commercial basis. However, improved bulls for natural breeding could be distributed free of cost by the government for the benefit of poor farmers. Rearing of such bulls will be the responsibility of Panchayat / cooperative societies / NGO's.

Field AI network (A.I. outlets), sperm stations, breeding farms and breeding programmes (Performance Recording, Progeny Testing, ONBS etc.) should constitute focal points for monitoring efficiency and progress.

Rapid computerization of the breeding network needs to be done in order to build up a reliable database and effective monitoring through a Management Information System (MIS) both at State and national Level.

Under the prevalent conditions in the country, the conventional method of producing progeny tested bulls has failed to achieve the desired results. Advance technologies like ETT and OPU-IVF should be used to support this programme.

Infrastructure For Cattle Development and Breeding Central Herd Registration Scheme

The Government of India, through the Department of Animal Husbandry, is running this Scheme, which envisages the registration of elite cows and buffaloes of breeds of national importance. The Scheme awards incentives for rearing of elite cows and male calves and provides a superior quality germplasm for superior breeding.

The Herd Book is a list of each breed, with milk production records and breed characteristics. Through a process of certification, elite breeds of cattle are identified for further breeding on a large scale, resulting in breed multiplication of superior stocks. The Scheme has a significant role in assisting the State Departments of Animal Husbandry, private sector players and Government undertakings in the procurement of elite dairy cows and buffaloes and using their progeny of high genetic potential for use in the cattle development programmes.

The main objectives of the Central Herd Registration Scheme are:

- To locate superior germ plasm of indigenous breeds in breeding tracts.

- To introduce milk recording for further propagation.
- To regulate sale and purchase within the country and abroad.
- To propagate and awaken consciousness among the breeders for scientific breeding, development and preservation of cattle, which would improve their (breeders') socio-economic conditions.

The CHRS Units are located in Rohtak (Haryana), Ahmedabad (Gujarat), Ajmer (Rajasthan) and Ongole (Andhra Pradesh). The indigenous cattle breeds covered by the Scheme are Gir, Kankrej, Haryana and Ongole.

Cattle Breeding Farms

Various breeds of national importance are being conserved at institutional and government farms in different parts of the country.

Central Cattle Breeding Farms

For rearing bull mothers of different breeds, seven Central Cattle Breeding Farms (CCBF) were established from 1967 to 1975 at Suratgarh (Tharparker and its crosses with Holstein Friesian), Chiplima (Red Sindhi and its crosses with Jersey), Sunabeda (Jersey), Andeshnagar (Holstein Friesian X Tharparkar), Hesserghatta (Holstein Friesian), Dhamrod (Surti) and Alamadhi (Murrah).

The primary objective of the farms was to produce at least 10 progeny-tested bulls in each farm by maintaining about 300 breedable females. This objective was never achieved and the programme to produce progeny tested bull was abandoned in 1988. The other objectives like genetic improvement of bull mothers of important cattle and buffalo breed and supply of high pedigree bulls also failed to achieve the target. The Working Group on Animal Husbandry for the Tenth Plan has suggested that the goals of these farms should be changed and they should be used either for conservation of indigenous breeds which are at the verge of extinction or NDDB could take over these farms for implementing progeny testing programme using recent technologies like ETT /OPU-IVF.

Central Frozen Semen Production

Central Frozen Semen Production and Training Institute established in 1969 at Hesserghatta is a premier Institute producing above 9 lakh doses of semen per year and imparting training to the field officers and veterinarians.

National Project for Cattle and Buffalo Breeding (NPCBB)

The Project came into being in October 2002 by merging two Centrally-sponsored Plan Schemes viz. i) Extension of Frozen Semen Technology and Progeny Testing Programme and ii) National BullProductionProgramme.The salient features of the Project are:

- The National Project on Cattle and Buffalo Breeding is expected to facilitate delivery of vastly improved artificial insemination (AI) services at the farmer's doorstep.
- It is envisaged that all breedable females among cattle and buffalo population, would be brought under organized breeding within a period of 10 years.
- Genetic qualities and availability of indigenous breeds will be improved and important indigenous breeds will be preserved.
- 14000 AI practitioners are expected to be gainfully employed as a result of the project.

Cattle Breeding Policy

According to the Working Group on Animal Husbandry for the Tenth Plan, the broad frame-work of cattle and buffalo breeding policy recommended for the country since the mid-sixties envisaged selective breeding of indigenous breeds in their breeding tracts and use of such improved breeds for upgrading of the non-descript stock. While the framework was accepted by the States, appropriate operationalisation of the same through field level programmes could not be done because of various reasons. Lack of interest in promoting Breed Organization / Societies and related farmers' bodies contributed to gradual deterioration of indigenous breeds. Majority of owners having indigenous breeds were not willing to accept AI, which was the major Government intervention for breed improvement. Eventually, the availability of good quality bulls needed for natural mating in the breeding tracts became scarce, leading to further deterioration of indigenous breeds' inthesetracts.

The Working Group further observes that large deviation from the laid breeding policy has occurred, which is quite obvious from the fact that crossbreeding which was to be taken up in a restricted manner and in areas of low producing cattle has now spread indiscriminately all over the country including in the breeding tracts of some of the established indigenous cattle breeds. Keeping In view current concerns for sustainability, maintaining environment and bio-diversity and conservation of energy, there is a rethinking on the development and use of indigenous breeds for milk and draught. The country since then has advanced in the area of newer reproductive technologies, which can be of tremendous advantage for rapid multiplication of elite germplasm. Therefore, a fresh look at the breeding policy is needed. The policy needs to be dynamic and consider, inter-alia, the demand for milk, requirement of draught animal power for agricultural and transportation purposes, need to conserve breeds in their breeding tracts, farming systems, production environments and availability of inputs as well as marketing channels. If uch a policy does not exist, the same has to be evolved and followed consistently for a reasonable period, say twenty years, after which the policy may be reviewed.

As regards Breeding Strategies, the Working Group felt that programmes for genetic improvement undertaken in the past were not successful, particularly those relying on up-gradation of indigenous breed through continuous crossbreeding for lack of backup support in the feed and fodder resources. However, in States like Punjab, where average milk production of 2500 litre per lactation in crossbred cows can be achieved under field conditions, substantial progressbeenmade.

Since most of the female stock is needed for herd replacement, accurate selection of sires assumes greater importance. But a feasible cost effective and proven method for general adoption in the country is yet to emerge. Any programme for genetic improvement needs an organization/set-up that goes beyond the individual/herd. Absence of breeders' organizations and field recording network are serious handicaps in the emergence of viable and effective Breeding Service Organizations. Genetic measures undertaken to improve livestock will not be successful unless the livestock production system as a whole is considered. Availability of inputs and support services, marketing channels and economic viability will have to be considered as an important component of the whole system. Rapid genetic changes in livestock population for efficient commercial production will have to be brought about by a carefully planned and monitored process. Conditions congenial to private initiatives to aid the process for faster improvement in productivity will assume paramount importance because the central and state governments may not be in a position to provide financial support for programmes in the long run.

The Working Group further states that the efforts would need greater attention because breeding is a cost intensive, long-term exercise with a time horizon of 15 years in India. Unless those who undertake such breeding programmes do not have a full control over various facets involved for this period, they run the risk of wasting time, effort and resources. If livestock development sector is to be successful, in terms of generating income to farmers, returns to government expenditures, and in value addition in international prices, the focus of policy will have to shift from the "best" technology to the most productive technology that is appropriate for different regions and is in tune with their natural endowment and labour and capital resources. Further, the adoption of appropriate breeding programmes and technology will result in accumulation of comprehensive field data on farmers' preferences, productivity of animals, cost of feed and other inputs, animal responses to nutrition, and other similar biological factors. A major systematic effort in this direction is required if an all round sustainable genetic improvement of cattle and buffalo is to be effected in the country. A reorientation of cattle & buffalo breeding policy would be attempted with area specific approach backed up by appropriate programs addressing our concerns for indigenous cattle breeds and draught animal power. A similar approach has been adopted in the National Project on Cattle and Buffalo Breeding.

Indigenous cattle breeds accepted by common farmers shall be further developed through region specific and breed specific programs aimed at selection in the breeding tracts and supply of improved quality germplasm on demand by farmers.

The states shall be directed to specifically delineate the areas of native breeds of cattle, record their numbers breed wise and sex wise and encourage farmers to conserve them in their home tracts.

Formation of breed associations for improvement of indigenous breeds shall be encouraged. Such associations shall be involved in production of quality male stock. An effective mechanism for providing disease free quality breeding bulls and quality semen for artificial insemination will be put in place. Breeding services would be provided at the farmers' door.

For sheep breeding also an area specific approach shall be adopted for effecting qualitative and quantitative improvement in carpet and coarse wool and developing fine wool. Breeding of sheep and goats will aim at increasing body weight, reproductive efficiency and control of mortality, besides improvement in milk yield in goats. Main focus will be on selection of rams and bucks and their distribution with backup by suitable programs.

In high altitude areas support for breeding of Yaks and mithuns shall continue.

Breeding of rabbits for fur and broiler purpose shall be encouraged in suitable areas.

Preservation and development of pack animals – horses, mules, donkeys and camels - shall also be considered.

The indigenous breeds of livestock and poultry are essentially the products of long term natural selection and are better adopted to withstand tropical diseases and perform under low and medium input. Many of these breeds may have useful genes for fast growth, prolificacy and small size. Such utility genes and breeds shall be identified, conserved and utilized. In recent time, international actions have been oriented towards conservation of animal genetic resources. India being a signatory to many such international agreements, the country will have specific policy focus on conservation of indigenous breeds of livestock and poultry.

Breeding Policy of India

Animal husbandry programmes have been run through the State schemes. Each State has to evolve its own breeding policy deciding on choice of breed, cross breeding strategy, optional mixture of animals of different breeds required, breeding goals in terms of expected genetic progress to be achieved, specific breeding programmes and the control measures that should be adopted to achieve the desired genetic gains in the population.

General parameters in the breeding policy formulated by various States are:

- Indigenous milch breeds such as Shaiwal, Red Sindhi and Gir, should be selectively developed for dairy traits in their native tracts.
- Indigenous dual purpose breeds such as Hariana, Tharparkar, Rathi, Kankrej, Gaolao, Ongole Deoni etc. should be developed selectively in their native tracts for dairy and draft traits.

- Indigenous draft breeds like Kangayam, Hallikar, Khillari, Amrit Mahal etc. should be developed selectively for draft traits in their native tract.
- Non-descript cattle will be bred with exotic semen to produce cross breed with Holstein Friesian or jersey and maintaing 50% exotic impenitence. In some States Red Sindhi, Tharparkar and Hariana have also been used upgrading non-descript cattle.

Development of Indigenous Breeds

To develop indigenous breeds Government of India has initiated three schemes namely National Project for Cattle and Buffalo Breeding, Central Herd registration scheme, Central Cattle Breeding Farms.

Central Herd Registration Scheme

For identification and location of superior germplasms of cattle and buffaloes, propagation of superior genetic stock, regulating sale and purchase, help in formation of breeders societies and to meet requirement of indigenous bulls in the different parts of the country. Government of India has initiated Central Herd Registration Scheme. Four CHRS units were established in different breeding tracts of the country. For milk recording 103 milk recording centers were set up. Indigenous cattle breeds covered under the scheme are Gir, Kankrej, Hariana and Ongole.

National Project for Cattle and Buffalo Breeding

Genetic improvement is a long term activity and Government of India has initiated a major programme from October 2000 "National Project for Cattle and Buffalo Breeding"(NPCBB) over a period of ten years, in two phases each of five years, with an allocation of Rs 402 crore for the 1stphase. National Project for Cattle and Buffalo Breeding envisages genetic up gradation on priority basis and also had focus on the development of indigenous breeds.

- The National Project for Cattle and Buffalo Breeding envisages 100 per cent grant in aid to implementing agencies and has the major objectives of
- To arrange delivery of vastly improved artificial insemination service at the farmers doorstep;

- To progressively bring under organized breeding through artificial insemination or natural service by high quality bulls, all breedable females among cattle and buffalo within a period of 10 years;
- To undertake breed improvement programme for indigenous cattle and buffalo breeds so as to improve their genetic qualities as well as their availability and
- To provide quality breeding inputs in breeding tracts of important indigenous breeds so as to prevent the breeds from deterioration and extinction.
- The project components specially designed to address the existing inadequacies will focus on the hitherto neglected natural meting system as well as the A.I. network with particular attention to
- Streamlining storage and supply of Liquid Nitrogen by sourcing supply from industrial gas manufacturers and setting up bulk transport and storage systems for the same,
- Introduction of quality bulls with high genetic merit,
- Promotion of private mobile A.I. service for doorstep deliver of A.I.,
- Conversion of existing stationery government centres into mobiles centres,
- Quality control of bulls and services at sperm stations, semen banks and training institutions,
- Study of breeding systems in areas out of reach of A.I. and
- Institutional restructuring by way of entrusting the job of managing production and supply of genetic inputs as well as Liquid Nitrogen to a specialized autonomous and professional State Implementing Agency.

CATTLE REEDING POLICY IN DIFFEENT STATES

S.No	State/UT	Breed	Breeding Policy
1.	Andhra Pradesh	Ongole	Selective breeding in Ongole: grading up, non-descript with Ongole
		Malvi	Selective breeding Malvi in pockets, grading of Malvi with Tharparkar and Deoni
		Hallikar	Selective breeding in Hallikar; grading up of nondescript with Hallikar
		Non-descript	Grading up, with Ongole, Tharparkar and Deoni cross breeding with Jersy and Holstein
2.	Arunachal Pradesh	Local cattle	Grading up, with Hariana and Redsindhi cross breeding with Jersy
3.	Assam	Local cattle	Grading up, with Hariana and Redsindhi; cross breeding with Jersy zc
4.	Bihar	Local cattle	Grading up, with Tharparkar Hariana and Redsindhi; cross breeding with Jersy
5.	Chattisgarh	Local cattle	Grading up, with Tharparkar Hariana and Shaiwal; cross breeding with Jersy and Holstein
6.	Gujarat	Gir, Kankrej	Selective breeding in Gir and Kankrej; grading up, non-descript with Gir and Kankrej; cross breeding with Jersy and Holstein-Friesian
7.	Goa	Local cattle	Grading up, with Redsindhi; cross breeding with Jersy
		Hariana	Selective breeding
		Sahiwal	Selective breeding
8.	Haryana	Non-descripit	Grading Up, non-descript with Hariana, Shaiwal, Tharparkar; cross breeding with Jersy and Holstein-Friesian.
9.	Himachal Pradesh	Local cattle	Grading up, with Hariana and Redsindhi; cross breeding with Jersy
10.	Jammu & Kashmir	Local cattle	Grading up, with Hariana and Redsindhi; cross breeding with Jersy
11.	Jharkhand	Local cattle	Grading up, with Tharparkar Hariana and Redsindhi; cross breeding with Jersy
		Deoni	Selective breeding
		Krishna Valley	Selective breeding
		Khillari	Selective breeding
12.	Karnataka	Amrit Mahal	Selective breeding
		Hallikar	Selective breeding
		Non-Descript	Grading Up, non-descript with Redsindhi; cross breeding with Jersy and Holstein-Friesian.
		Local cattle	Grading Up, non-descript with Redsindhi, Kangayam and Tharparkar; cross breeding with Jersy and Holstein-Friesian.
13.	Kerala	Crossbreds	Selective breeding with F1 cross bred bulls

			obtained from progeny tested either jersy or Holstein bulls
		Nimari	Selective breeding
		Malvi	Selective breeding
14.	Madhya Pradesh	Kenkatha	Selective breeding
		Non-descript	Grading up, with Gir, Tharparkar, Hariana Sahiwal and Ongole; cross breeding with Jersy and Holstein
		Khillari	Selective breeding
		Dangi	Selective breeding
		Gaolao	Selective breeding
15.	Maharashtra	Nimari	Selective breeding
		Non-descript	Grading up, with the breeds of the region and Hariana; cross breeding with Jersy and Holstein
16.	Manipur	Local cattle	Grading up, with Redsindhi; cross breeding with Jersy
17.	Meghalaya	Local cattle	Grading up, with Redsindhi; cross breeding with Jersy
18.	Mizoram	Local cattle	Grading up, with Hariana; cross breeding with Jersy
19.	Nagaland	Local cattle	Grading up, with Hariana; cross breeding with Jersy
20.	Orrisa	Local cattle	Grading up, with Redsindhi and Hariana; cross breeding with Jersy and Holstein
21.	Punjab	Local cattle	Grading up, with Sahiwal and Hariana; cross breeding with Holstein Friesian and Jersy
		Nagori	Selective breeding
		Malvi	Selective breeding
		Rathi	Selective breeding
22.	Rajasthan	Non-descript	Grading up, with Hariana, Gir, Tharparkar and Rathi; cross breeding with Jersy and Holstein Friesian
		Siri	Selective breeding
23.	Sikkim	Local cattle	Grading up, with Hariana; cross breeding with Jersy
		Kangayam	Selective breeding
		Umblachery	Selective breeding
24.	Tamilnadu	Bargur	Selective breeding
		Non-descript	Grading up, with Red sindhi; cross breeding with Jersy Holstein Friesian
25.	Tripura	Local cattle	Grading up, with Tharparkar; cross breeding with Jersy
		Kenkatha	Selective breeding
26.	Uttar pradesh	Non-descript	Grading up, with Hariana, Sahiwal, Tharparkar and Redsindhi; cross breeding with Jersy and Holstein Friesian

Breeding Policy for Uttarakhand State

Strategies for implementing the breeding programme include

A.	Upto 1000 meter Altitude (Tropical Zone)	1.	Crossbreeding with H. F. and jersey semem will be provided to those who prefer Jersey.
		2.	Interse mating in crossbreds or Breeding of known indigenous breedy by DFS of the same breed.
		3.	Breeding of known indigenous breeds.
B.	1000-1500 meter Altitude (Sub tropical)	1.	Crossbreeding of local cattle with Jersey and H. F. semen will be provided to those who prefer H. F.
		2.	Intese mating of F_1 with Half bred.
		3.	Breeding of known indigenous breeds by DFS
C.	1500-2400 meter Altitude (Cool Temprature Zone)	1.	Selective Breeding in local nondescript Cattle.
		2.	Breeding of known indigenous breeds by DFS of the same breed.
D.	Above 2400 meters Sub alpine/Alpine Zone	1.	Selective breeding among the local nondescript cattle.

Establishing an institutionalized monitoring mechanism from the centre down words.

Drawing of region specific and breed specific breeding strategies in each state.

Fixing targets in terms of actual members of cattle sheds of the particular breed, infrastructure facilities such as spermstation, bull farms etc. for each state.

Organizing regular review meeting to asses the results in terms of the physical and financial targets.

Ensuring adequate funding from central Government to the state Government to implement the programmes.

Poultry Breeding Policy and Objectives:

To foster the cause of holistic development for the rural and socially backward communities, gender empowerment, poverty alleviation and nutritional nourishment, especially of the rural poor, using poultry as a tool.

To increase per capita availability of poultry products by stepping up sustainable production using suitable technology.

To build participatory institutions of collective actions for small scale farmers this will allow them to vertically integrate with input suppliers and processors.

To crease an environment through training and extension to enable farmers to adopt technology for improving productivity and management of the birds.

To promote creation of infrastructure like laboratories etc. and boost processing and distribution facilities along with commensurate export promotion to enable increase in exports.

To promote effective regulatory institutions for dealing with environmental and health crises arising out of poultry and poultry products.

Breeding Policy

India is recognized as mega diversity zone and it is necessary foremost that the seed or the stock is chosen, suitable to the agro-climatic condition of the region. Impetus is needed for the development and adoption for the low input technology birds suitable for training at village conditions. To survive at the farmers' door, natural refractoriness to common diseases should also be an attribute of the birds. The States should get the evaluation of the performance of the existing stocks done and replenish them wherever needed and phase out wherever not. The breeding policy for such State requires specific attention to the local birds.

The indigenous birds have been researched in most of the regions and based on their production, adaptability to the environment, disease resistance; selection should be made for their propagation. The indigenous breeds are like genetic insurance, which may be needed in future and is also closer to the masses due to sentimental value. The diversification of poultry by taking up breeding of ducks, Japanese quails, turkey and guinea fowl will be given thrust. The inputs and marketing facilities for these shall be geared up prior to ensuring their popularization. Ostrich and Emu shall also be considered for introduction on commercial scale, keeping in view the environmental implications.

Establishment of a gene pool or bank for indigenous breeds and threatened breeds as a conservation strategy will be considered. Studies and conservation of Red Jungle Fowl, the progenitor of modern day fowl will be undertaken as its native tract is believed to be the Indian sub continent and the knowledge may be used to our advantage in future.

An "Avian Genetic Resource Task Force: shall be set up to popularize propagation of already recognized indigenous breeds like Aseel, Ankaleshwar, Buser, Brown desi, Chatting (Malay), Daothigir, Denki, Frizzle, Phages, Haringhata, Kalasthi, Faverolla, Naked neck, Punjab Brown, Tellicherry, Titni, Teni, Nicobari and duck breeds like Indian Runner, Nageshwari, sythetmete etc.

The nutritive/special characteristics of the poultry products from indigenous birds, if any, shall be scientifically validated and documented for wide publicity. In the commercial sector, nearly 15% of the production is dependent on production of stocks from imported grandparent stocks, certain lines of which have to be continually imported. The safety of import of these stocks in terms of prevention of ingress of exotic diseases shall be ensured, through compliance of sanitary requirements. There are presently no benchmarks regarding the field performances of the improved indigenous low input technology stocks, may be because other attributes like colored plumage, tinted eggs etc. complement for lower production. However, a threshold field level output will be studied and documented.

Besides laying stress on traits for improvement in breeding program, additional traits like improvement in carcass quality and dressing yield for broilers along with improvement in egg mass and egg quality characteristics for layers should also be considered by the breeders, in the commercial sector.

Chapter - 22

Conservation of Genetic Resources in India

Introduction

India has vast animal genetic resources with a wide variety of indigenous farm animals including cattle. The cattle breeds have evolved over generations to adapt to the agro-climatic and socio-economic needs of the people. Domestic animal diversity is defined as the spectrum of genetic differences within each breed and across all breeds within each domestic animal species, together with the species differences; all of which are available for the sustainable intensification of food and agriculture production. The domestic animal diversity has evolved over millions of years through the processes of natural selection forming and stabilizing each of the species used in food and agriculture. Over the more recent millennia the interaction between environmental and human selection has led to the development of genetically distinct breeds. Selection processes, directed by both humans and the environment, together with the random sampling processes causing genetic populations to drift over generations, have accelerated the development of the diversity within species leading to the creation of distinct genetic differences amongst breeds.

Genetic Diversity in India

Species	As per World Watch List	As per Indian Literature
Cattle	70	30
Buffalo	20	10
Yak	5	Nil
Sheep	62	42
Goat	34	20
Horse	7	6
Donkey	3	Nil
Camel	9 (+1 Bactrian)	8
Pig	8	Nil
Rabbit	3	Nil
Fowl	19	15
Quail	2	Nil
Duck	6	Nil

India is the seventh largest country in the world and it is recognized as one of the 12-mega biodiversity centres of the world. It is well marked off from the rest of Asia by mountains and the sea, which gives the country a distinct geographical entity. Due to diverse agro-ecological regions and topographic conditions, India has rich repository of both flora and fauna. India has vast animal genetic resources with a wide variety of indigenous farm animals. It has 132 registered farm animal breeds viz. 30 cattle, 10 buffaloes, 42 sheep, 20 goats, 7 camel, 5 horses/ponies and 18 chickens, besides many other non-descript and mixed populations. Livestock husbandry is an age-old important occupation for Indian farmers. The unique and rich animal biodiversity is extensively referred to in Indian scriptures.

Justification For Conservation or Reason?

Different reasons for conservation of animal genetic resources include-

1. Economic and Biological Reasons

A. Genetic variation both within and between breeds is the raw material with which the animal breeder works. Therefore, any loss of genetic variation will limit our capacity to respond to changes in economic forces for the exploitation of animal production in future.

B. Breeds with specific qualities like disease resistance, heat tolerance, ability to survive and produce under stress and low input conditions need to be preserved for future use.

C. Future requirements of type and quality of animal produce (milk, draught power) may change and this requires conservation of animals with better performance in specific production traits.

D. Magnitude of heterosis depends upon the breeds crossed. For exploiting the heterosis in animal production, it is necessary to maintain breeds which are complementary to each other and on crossing result in maximum heterosis.

2. Scientific Reasons

A. Breeds with unique physiological or other traits are of great value as they provide missing links in the genetic history of a livestock species by the study of blood groups or polymorphic traits. To identify the DNA sequences causing the distinctive traits, preservation of breeds with unique traits will be essential for long term research in molecular engineering.

B. To evaluate the magnitude of genetic change due to selection, maintenance of a sample as control population is very much essential.

C. Investigations in different areas like physiology, biochemistry, genetics immunology, etc. Require maintenance of diverse populations.

D. Variety of populations are an asset for research work in biological evolution, behavioural studies, etc.

E. Diverse populations form an excellent teaching material for students of animal science, ecology, etc.

3. Historical and Cultural Reasons

Conservation of historically important, culturally interesting and visually unusual and attractive population is very important for education, tourism etc. Further, conservation of breeds

A. Can be a valuable material of nature and culture,

B. Serve as research and teaching material in history and

ethnography,

C. Will be preservation of populations with diverse sizes, colours and other morphological features, for aesthetic reasons, and

D. Need be done to take care of existence of different creations of the nature for posterity.

Status of Species -Female Population

S. No.	Status	Number
1	Normal	>10,000
2	Insecure	10,000-5,000
3	Vulnerable	1,000-5,000
4	Endangered	100-1,000
5	Critical	< 100

Mechanism of Conserving Cattle Genetic Resources

Once genetic resources have been identified and characterized, two basic conservation activities can be followed,

1. *In Situ* Conservation
2. *Ex-Situ* Conservation

In Situ Conservation

1.	Haploid forms	a.	Frozen semen
		b.	Frozen eggs/oocyte
2.	Diploid forms	a.	Frozen embyo
		b.	Live animal

In situ conservation requires establishment of live animal breeding farms and their maintenance. The generation and loss of alleles is a dynamic process that should be maintained at close equilibrium through sound management. In situ conservation strategies emphasize wise use of indigenous cattle genetic resources by establishing and implementing breeding goals and strategies for animal sustainable production systems. Information for animal recording and breeding is well established in developed countries through breeding associations which zealously protect the interest of breeds including rare ones.

Infrastructure appropriate to systems in developing countries remains scarce. In India, such efforts are limited to only six breeds of cattle, in a herd registration program organized by the Central Animal Husbandry Department. Similar programs are required for the rest of the breeds and species as well. In any such program, the success depends upon the participation of the farmer for which he needs support and incentive. Therefore, it is difficult to organize the farmers for conserving the breeds which are no more economical to him. In the case of breeds which are no more economically viable, therefore, the only alternative is to bring them under government farms. In situ conservation is very costly if the entire population has to be retained for which at least 26 females and 10 males in cattle have to be maintained that would keep the inbreeding coefficient at 0.2 per cent per year. Therefore, this approach would have to be limited to those breeds which are highly endangered. Modalities for simplified animal recording, genetic development and dissemination are needed for each species for a range of national livestock structures in developing countries.

Major Advantages of In-Situ Conservation

1. Live animals can be evaluated and improved over the years.
2. Genetic defects can be detected and eliminated.
3. Live animals are always available for immediate use.
4. They are a gene bank for future use.
5. They are a constant reminder that the needs of posterity must be considered.
6. The herd may have some economic advantages (heat tolerance, disease resistance) which can be exploited and so render the enterprise economically viable.
7. The produce from live animals partly compensates the expenditure, if not entirely.
8. From aesthetic point of view, the live animals are, visible, a pleasure to look at, the people are delighted to see variety of animals and have some cultural value.

The Major Limitation of Live Animal Conservation

While fixing the number for preservation of a breed,

1. The cost of maintenance,
2. Availability of animals and rate of inbreeding should be taken into consideration.
3. With small population size, the effective population size decreases and the genetic structure of the population is affected due to inbreeding and random drift.
4. Many models are now available which reduce inbreeding to a minimum, but random drift over long periods may lead to a population very different in genetic composition from the initial one.
5. Gene X environment interactions is another disadvantage.
6. In situ conservation involves
   - A large infrastructure of land,
   - Buildings,
   - Feed and fodder resources,
   - Water supply,
   - Labour,
   - Technical and supervisory man-power, etc.

It should be made to maintain a rate of inbreeding of less that 0.5% per generation for long term programmes while slightly higher rate could be tolerated for short term programmes. A flock/herd of 150 breeding females with 20 breeding males, the males being unrelated as far as possible, for preservation purposes. A maximum inbreeding level of 1' % per generation as tolerable and a herd size of 200 breeding animals in necessary to breed and selected successfully for a quantitative trait.

Ex-Situ Conservation

Ex-situ conservation includes cryogenic preservation. It is the storage of genetic resources, which the farmers are currently not interested in using. Ex situ conservation is based on the use of live animals populations wherever practicable, supported by cryopreservation where technology exists or can be developed, combining within-country gene banks with global repositories. Interested governments, nongovernmental organizations, research institutions and private enterprises should be encouraged to maintain in vivo samples of breeds at risk, with national inventories being established and kept up to date so that the genetic resources are readily available for use and study. Because of random drift and possible gene by environment interactions, ex situ methods are generally preferred over in situ. Ex situ conservation is comparatively more convenient, economical and easy with the application of modern reproductive technologies.

Advantages

1. If the preservation is to maintain populations without genetic change, it can be best done by cryogenic storage as it is difficult to breed many generations of animals without any change in the genetic structure.
2. The resources requirement for in situ preservation is quite large as compared to cryogenic methods.

Limitations

1. Ex situ preservation using frozen semen delays the restoration of a breed as it can be restored in the future only by upgrading. But this could be overcome through preservation of embryos.
2. Another important factor is the danger faced by a breed restored from cryogenic preservation from important changes in the environment like germs, climate, etc., that have taken place over the years.
3. Variability in cryogenic storage of germplasm, accessibility to their physical location, ownership, behaviour of animal, response of germplasm to freezing and thawing techniques, and poor conception rate

Ex-Situ/Cryogenic Preservation Includes

1. Preservation of frozen semen
2. Preservation of oocytes
3. Preservation of embryos
4. Preservation of ovaries
5. Use of embryonic stem cells or blastomeres
6. Production of chimeras

7. Production of embryos in vitro
8. Embryo splitting
9. Transgenesis
10. DNA libraries

1. Semen Preservation

Semen cryopreservation and artificial insemination are important tools of animal improvement with vast scope of genetic improvement, conserving indigenous cattle resources because of their simplicity and relatively cheaper costs. Semen storage and distribution activities are being carried out in a few well-known indigenous milch cattle. Still, many breeds like Amrithmahal, Dangi, Gaolao, and Punganur lack such facilities. Spermatozoa may also be collected from the epidydimis

2. Oocyte Preservation

This method provides an opportunity for conserving females in the same way as sperm is conserved. The oocyte can be recovered by surgery, laparotomy or slaughtering of donor animals. The frozen thawed oocyte can be used for IVF successfully. The immature and mature oocyte from slaughtered animals could be useful in near future for cryopreservation of genetic material of endangered breeds.

3. Embryo Preservation

The main advantage of cryopreservation of embryo over that of sperm or oocyte is that it contains the complete genome. The embryo transfer technique coupled with micro manipulation, embryo sexing and splitting are more useful and economical to carry out ex situ conservation of animal genetic resources. This technique is very useful in conservation of genetic resources by rapid multiplication of superior or rare germplasm. MOET can be used in resurrecting the endangered cattle breeds like Sahiwal, Punganur and Vechur.

4. Preserving of Ovaries

The preservation of slices of ovaries in liquid nitrogen is a new technology which may be of great use in conservation of animal genetic resources. The ovary slices might be transferable into suitable recipients to obtain oocytes which can be fertilized.

5. Embryonic Stem Cell and Nuclei

Preservation of embryonic stem cells could represent an important method of genome conservation and would be helpful in propagating animals from a single embryo of elite or rare animals belonging to endangered breed/species. Embryonic stem cells represent progressively growing cultures of embryonic cells which retain their pluripotential characteristics. They are derived by culturing blastocysts in vitro in such a way that the cells from the inner cell mass proliferate but do not differentiate. Embryonic stem cell lines can be isolated and then multiplied by continued culture. The importance of embryonic stem cell lines is that, if they are incorporated with normally developing embryos, they will participate in the formation of the inner cell mass and produce chimaeric animals, including germ line chimaeras which are fertile. A more direct route of regenerating animals from embryonic stem cells might be to use the nuclei from these for nuclear transplantation into enucleated oocyte and embryonic multiplication.

6. Chimaeras

Chimaeras mean the animals having body cell population with different karyotypes which have been formed from two or more zygotes with different karyotypes. Chimeric embryos have been frequently made by the aggregation of cells from two individual embryos or by injecting cells from one embryo into the blastocyst cavity of another embryo. So long as cells from both embryos are represented in the inner cell mass, the composite embryo will develop into a chimaeric animal. A more important aspect of chimaerism in genetic conservation is the potential use for inter-species embryo transfer. Using this technology sheep have been born to goat foster mother and vice versa. This would be especially important if it was a species rather than a breed of animal that was on the verge of extinction.

7. Production of Embryos in Vitro

This technology involves salvaging mature oocyte from ovaries of slaughtered animals and developing methods for their maturation, fertilization and in vitro culture for normal embryonic development. For conservation of rare breeds, such a method could provide an opportunity to salvage a few oocytes from the ovaries of rare and superior animals even after their normal reproductive life and to produce some blastocysts for conservation by deep freezing.

8. Embryo Splitting

Embryo splitting is more advantageous in the circumstances where only few embryos of particular genotype or breed or species are available. This technique of manipulating embryos could be helpful in producing more number of animals from a few stored embryos of rare and endangered animals.

9. Transgenesis

Recent developments in molecular biology have enabled introduction of specific genes into the animal genome. A small amount of DNA is injected into the nucleus of an egg soon after fertilization. In some instances the DNA becomes integrated into the chromosomes and an embryo and foetus develops in which all the nuclei contain copies of the inserted gene. As any sequence can be spliced with any other (Anderson, 1986), transportation of genes across breeds and species is possible by recombinant DNA techniques. Successful microinjection of genes in mice embryo and their expression .the creation of transgenic forms a possibility, and for the future a viable technique for improving animal production, and for conservation and capitalize on of important genes across breeds and species.(Gordon et al)

10. DNA Libraries

Theoretically, an animal can be produced from its complete DNA complement. However, at present technical developments are limited to the identification and manipulation of only a few genes. But the direction in which technical advancements are taking place gives an indication that in future breeds/species can be reconstructed from their DNA complements. This gives the hope that if complete DNA complement is stored either in lyophilised form or as cryopreserved cells, reconstruction can be taken up when techniques are standardised. Advances in biotechnology are emerging in a big way and could be of great utility in near future for animal improvement and animal genetic resource conservation programmes.

The Main Causes of the Reduction in Numbers of Indigenous Livestock are-

1. The intensive efforts being made to introduce superior germplasm to overcome the low production of the local genotypes, without any consideration to the local ecological

and socio-economic situation (demand for increased milk production and introduction of exotic milk breeds of cattle like Jersey, Holstein).

2. No serious organized effort in maintaining the purity of the breeds - inter-mating among breeds located in each others' vicinity resulting in dilution of breed characteristics and Changes in cropping pattern and increased mechanization of agriculture with neglect of indigenous draught breeds of cattle.

3. The global awareness for conservation of domestic animal diversity was touched up on at the International Conference held in Rome in 1936, which resulted in several recommendations being made in this regard. Based on these recommendations, the Government of India had initiated herd registration schemes for several established breeds of cattle, viz., Gir, Hariana, Kankrej, Ongole and Tharparkar.

Standards of performance were laid down for registration of animals both in registered and farmers' herds, and also including milk recording of animals. However, this scheme was not entirely successful. Efforts were made from time to time by the Government of India/ Indian Council of Agricultural Research to define the breed characteristics of important breeds of livestock and a bulletin containing breed characteristics of important breeds of cattle and a buffalo was published by ICAR in 1979.compiled information on important indigenous breeds of livestock.

References

Berge, S., 1961. The historical development of animal breeding. In: Scientific problems of recording systems and breeding plans of domestic animals (E. Schilling, ed). 11, 109-127. Max-Planck. Inst. Tierz.Tierernährung, Mariensee/Trenthorst.

De Vries, A. G., 1989. A method to incorporate competitive position in the breeding goal. Anim. Prod. 48:221-227.

Falconer, D. S., 1960. Introduction to Quantitative Genetics. Oliver and Boyd, London.

Falconer, D. S., and T. F. C. Mackay, 1996. Introduction to Quantitative Genetics. Longman, Harlow (3rd edition).

Hazel, L. N., 1943. The genetic basis for constructing selection indexes. Genetics 28:476-490.

Hazel, L. N., and J. L. Lush, 1942. Efficiency of three methods of selection. J. Hered. 33:393-399.

Henderson, C. R., 1952. Specific and general combining ability. In: Heterosis (J.W. Gowen, ed). pp. 352-370. Iowa State College Press, Ames, Iowa. Henderson, C. R. 1950. Estimation of genetic parameters.Ann. Math. Statist. 21:309.

Henderson, C. R., H. W. Carter and J. T. Godfrey, 1954. Use of contemporary herd average in appraising progeny tests of dairy bulls. J. Anim.Sci. 13:949.

Hopkins, I.R., 1978. Some optimum age structures and selection methods in open nucleusbreeding schemes with overlapping generations. Anim. Prod. 26: 267-276.

Hopkins, I.R. and James, J.W., 1978. Theory of nucleolus breeding Schemes into overlappinggeneration. Theoretical and applied genetics, 53:17-24 .

James, J.W., 1977. Open nucleus breeding systems. Animal Production. 24: 287-305.

James, J.W. 1978. Effective population size in open nucleus breeding scheme. Acta Agric. Scand. 28:387-392.

Jasiorowski H.A., 1990. Open nucleus breeding schemes - new challenge for the developing countries. In: Zurkowski M. (ed), Proceedings of a Conference on Open Nucleus BreedingSystems, Biatobrzegi, Poland, 1-19 January 1989. FAO (Food and Agriculture Organization of the United Nations), Rome, Italy. pp. 7-12.

Kasonta, J.S. and Nitter, G., 1990. Efficiency of nucleus breeding scheme in dual purposecattle in Tanzania. Animal Production 50: 295-251.

Keller D.S., Gearheart W.W. and Smith C., 1990. A comparison of factors reducing selection response in closed nucleus breeding schemes. Journal of Animal Science 68:1553-1561.

Lush, J. L., 1945. Animal Breeding Plans. Iowa State College Press, Ames, Iowa.- 34 -

Lush, J. L., 1994. The Genetics of Populations. Prepared for publication by A. B. Chapman, R. R. Shrode and withan addendum by J.F. Crow. College of Agriculture, Iowa State University, Ames, Iowa. Special Report 94.

Mueller, J.P. and James, J.W., 1983. Effects of reduced variance due to selection in opennucleus breeding systems. Australian Journal of Agricultural Research 34(1):53-62

Nicholas F.W. and Smith C., 1983. Increase rate of genetic change in dairy cattle by embryo transfer and splitting. Animal Production 36:341-353.

Shepherd, R.K., 1991. Multi-tier open nucleus breeding schemes. In: PhD. Thesis. Armidale.University of New England.

Rendel, J.M., and A. Robertson, 1950. Estimation of genetic gain in milk yield by selection in a closed herd of dairy cattle. J. Genet. 50:1-8.

Robertson, A., 1961. Inbreeding in selection programmes. Genet. Res. 2:189-194.

Smith, C., 1985. Scope for selecting many breeding stocks of possible economic value in the future. Anim. Prod.41:403 -412.

Smith C., 1988. Genetic improvement of livestock in developing countries using nucleusbreeding units. World Animal Review 6:2-10.

Thompson, R., 1986. Estimation of realized heritability in a selected population using mixed model methods.Génét. Sél. Evol. 18:475-484.

Verrier, E., J. J. Colleau, and J. L. Foulley, 1991. Methods for predicting response to selection in small populations under additive genetic models : a review. Livest. Prod. Sci. 29:93-114.

Willham, R. L., 1999. On Jay Lush. In: From Lush to Genomics: Visions for the Future of Animal Breeding and Genetics, Iowa State University, Ames, Iowa. May 16-18, 1999.

Wray, N. R., J. A. Woolliams, and R. Thompson, 1990. Methods of predicting rates of inbreeding in selected populations. Theor. Appl. Genet. 80:503-512.

Yapi C.V., Oya A. and Rege J.E.O., 1994. Evaluation of an open nucleus breeding programmefor growth of the Djallonke sheep in Côte d'Ivoire. In: Proceedings of the 5th WorldCongress on Genetics Applied to Livestock Production, Guelph, Canada, 7-12 August1994. Volume 20. International Committee for World Congress on Genetics Applied toLivestock Production, Guelph, Ontario, Canada. pp. 421-424.